Lecture Notes of the Institute for Computer Sciences, Social Informatics and Telecommunications Engineering 318

More information about this series at http://www.springer.com/series/8197

Paulo Pereira · Rita Ribeiro ·
Ivo Oliveira · Paulo Novais (Eds.)

Society with Future: Smart and Liveable Cities

First EAI International Conference, SC4Life 2019
Braga, Portugal, December 4–6, 2019
Proceedings

 Springer

Editors
Paulo Pereira
Centre for Territory, Environment
and Construction, CTAC
University of Minho
Guimarães, Portugal

Rita Ribeiro
Communication and Society
Research Centre, CECS
University of Minho
Braga, Portugal

Ivo Oliveira
Landscape, Heritage and Territory
Laboratory, Lab2PT
University of Minho
Guimarães, Portugal

Paulo Novais 🄳
Centro ALGORITMI
University of Minho
Braga, Portugal

ISSN 1867-8211   ISSN 1867-822X (electronic)
Lecture Notes of the Institute for Computer Sciences, Social Informatics
and Telecommunications Engineering
ISBN 978-3-030-45292-6   ISBN 978-3-030-45293-3   (eBook)
https://doi.org/10.1007/978-3-030-45293-3

This Springer imprint is published by the registered company Springer Nature Switzerland AG
The registered company address is: Gewerbestrasse 11, 6330 Cham, Switzerland

Preface

We are delighted to introduce the proceedings of the first edition of the European Alliance for Innovation (EAI) International Conference on Society with Future: Smart and Livable Cities (SC4Life 2019). This conference brought together researchers, developers, and practitioners from around the world who are leveraging and developing new knowledge on the topic of "smart cities", offering more efficiency to main infrastructures, utilities, and services, while creating a sustainable urban environment that improves the quality of life for its citizens and enhances economic development.

However, despite the accepted concept of smart cities, as a result of the utilization of the latest information and communication technologies (ICT), the concepts of smart, sustainable, resilient, safe, inclusive, and liveable cities, must be jointly considered, linking ICT advancements with the most innovative technology and systems of the built and natural environment, as well as with the most relevant component being its citizens, to offer efficient cities and territories. With this context in mind, the theme of SC4Life 2019 was "Society with a Future – Smart and Liveable Cities".

In this first edition, the following main topics were considered: Cities and Territories, Information and Technologies, and Citizen-centred Needs. In addition to papers in the referred topics, contributions with an interdisciplinary nature were particularly encouraged. The technical program of SC4Life 2019 consisted of 13 full papers, presented in oral sessions at the main conference tracks.

Apart from the high-quality technical paper presentations, the technical program also featured a keynote speech by Professor Filippo Practicó from the Università Mediterranea di Reggio Calabria, Italy.

Coordination with the general chair, Paulo Pereira, and the publication chair, Rita Ribeiro, was essential for the success of the conference. We sincerely appreciate their constant support and guidance. It was also a great pleasure to work with such an excellent Organizing Committee and we thank them for their hard work in organizing and supporting the conference. In addition to this, we would like to highlight the great work led by the Technical Program Committee, including Paulo Novais and Ivo Oliveira. We are also grateful to conference manager, Karolina Marcinova, for her support and to all the authors who submitted their papers to SC4Life 2019.

The SC4Life conference series provides a liveable forum for the exchange of scientific ideas and practical results, technological innovations and social challenges, and new visions, from academics, companies, and municipalities, also taking into account the irreplaceable role of the intervention of citizens, fostering the innovations to support a better quality of life with social cohesion.

December 2019   Paulo Pereira

Organization

Steering Committee

Imrich Chlamtac   University of Trento, Italy

Organizing Committee

General Chair

Paulo Pereira   University of Minho, Portugal

General Co-chair

Miguel de Castro Neto   NOVA University Lisbon, Portugal

TPC Chair and Co-chair

Paulo Novais   University of Minho, Portugal
Ivo Oliveira   University of Minho, Portugal

Publications Chair

Rita Ribeiro   University of Minho, Portugal

Web Chair

Karolina Marcinova   EAI

Panel Chairs

Paulo Pereira   University of Minho, Portugal
Paulo Novais   University of Minho, Portugal
Rita Ribeiro   University of Minho, Portugal

Technical Program Committee

Adriano Moreira   University of Minho, Portugal
Ana Melo   University of Minho, Portugal
António Pereira   Polytechnic of Leiria, Portugal
Catarina Sales   University of Beira Interior, Portugal
Cesar Analide   University of Minho, Portugal
Cristina Cavaco   University of Lisbon, Portugal
Daniel Casas Valle   Urban Dynamics, The Netherlands
Davide Rua Carneiro   Polytechnic of Porto, Portugal
Emília Araújo   University of Minho, Portugal

Contents

Citizen-Centre Needs

Cities and Territory

Sustainable Road Infrastructures Using Smart Materials, NDT, and FEM-Based Crack Prediction

Rosario Fedele[1]([⊠]) [iD], Filippo Giammaria Praticò[2] [iD], and Gianfranco Pellicano[2] [iD]

[1] CNR - Institute of Marine Engineering (INM), Via di Vallerano 139, Rome, RM, Italy
rosario.fedele@unirc.it

[2] DIIES, University "Mediterranea" of Reggio Calabria, Via Graziella, Reggio Calabria, Italy

Abstract. Smart cities need roads with high levels of sustainability. This goal can be reached using different approaches, such as smart materials, Non-Destructive Test (NDT)-based monitoring systems, and Finite Element Method (FEM)-based damage prediction models. The pieces of information provided using the above-mentioned approaches play a crucial role in the work of many stakeholders (citizens, users, road agencies, authorities, driverless vehicles, etc.). Consequently, the main objectives of this study presented in this paper are (i) providing an overview of the current approaches, and (ii) presenting a NDT-, and FEM-based monitoring system that was designed to improve the sustainability of the present and future road pavements by means of the road pavement damage detection and prediction. In more detail, the paper is focused on the set up and the calibration of a FEM model that aims at simulating the vibro-acoustic signatures of un-cracked and cracked road pavements. An NDT apparatus was used to gather the vibro-acoustic signatures of road pavement (data set) that was progressively damaged. Subsequently, the data set mentioned above was used to set up the FEM model. Results show that, even though the FEM model is able to replicate only in part the measured signals, this model can be successful used for predicting the variation of the structural health status of the road pavement. Hence, the proposed approach can be used to improve the sustainability of the current road pavements.

Keywords: Finite element modeling · Vibro-acoustic signature · Road crack prediction

1 Introduction

Smart cities require roads with high levels of sustainability, which can be reached using different approaches. Among all the possible solutions, the study presented in this paper is focused on three promising approaches (see Fig. 1), i.e., the use of smart materials, the implementation of Non-Destructive Test (NDT)- and sensor-based monitoring systems, and the application Finite Element Method (FEM)-based damage prediction models. In more detail: (i) smart materials can be used to obtain self-healing road pavements, to carry out energy harvesting, and to save resources (e.g., raw material, and

P. Pereira et al. (Eds.): SC4Life 2019, LNICST 318, pp. 3–14, 2020.
https://doi.org/10.1007/978-3-030-45293-3_1

energy); (ii) using NDT- and sensor-based monitoring system, instead of the traditional destructive ones (e.g., coring), can improve the sustainability of the roads because of the fact that it positively affects the management process (e.g., maintenance based on real time information about the road conditions); (iii) FEM models can be used to predict the temporal and spatial behavior of the road pavements, i.e., to forecast the occurrence of possible failures (e.g., internal cracks) or the reduction of the performances (e.g., clogging). For these reasons, the following subsections contain an overview of noteworthy applications of the three approaches listed above.

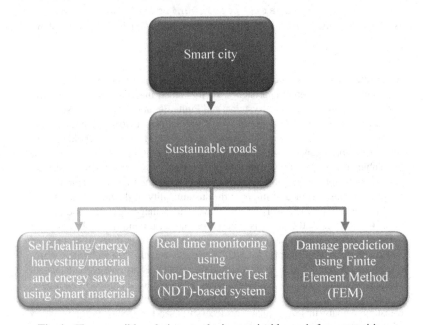

Fig. 1. Three possible solutions to obtain sustainable roads for smart cities.

1.1 Smart Material for Road Pavements

A material can be defined "smart" if it has the ability of changing its properties to react to an external condition [1]. In order to obtaining smarter, and more sustainable and efficient materials for road pavements, several types of wastes can be used in the asphalt concrete mixture, such as Reclaimed Asphalt Pavement (RAP) [15], by-products (e.g., crumb rubber, plastics, blast furnace slag, fly ash, leachate, glass, concrete, wood ash) [12, 14], graphene [7], and fibers [3, 8, 21]. These latter are used in applications that principally aim at conferring to the asphalt concrete healing properties (i.e., make the material able to easily repair the damages due to different causes, such as vehicular traffic, or thermal excursions) [4, 11, 16–19]. Furthermore, smart materials be used to produce or harvest energy from the road pavements. The energy derived from the roads can be used for typical applications, such as street lighting [22], and sensors based monitoring systems [5, 13], or for innovative applications, such as feeding electric vehicles [2, 9].

1.2 NDT Solutions for Monitoring the Road Pavements

Infrastructure monitoring can be carried out using [6] traditional destructive testing (DT), or innovative semi-destructive testing (SDT) and non-destructive testing (NDT). Usually, DT-based monitoring is the most used although it provides sample-based information (i.e., discrete points of the pavement), and requires energy and money for extracting (e.g., coring), analyzing, and landfilling pavement samples. On the other hand, SDT- and NDT-based methods, which are driven by the increasingly insistent demand for smart cities, are growing because they offer high performance (e.g., extended measurements), sustainability (e.g., energy and time savings), and efficiency (e.g., high measurement frequency, and/or technologically advanced devices). The main NDT-based monitoring drawbacks are related to the costs (i.e., instrumented infrastructures are more expensive than traditional ones), and to the worker's skill (i.e., skilled worker are required to set up, use, and tune sophisticated devices/systems, and/or to handle and analyze huge amounts of data). From an energy point of view, it should be underlined that the NDT methods tend to be more sustainable than the DT methods, because they use efficient and advanced systems (e.g., [10]).

Noteworthy NDT methods applied or designed for the Structural Health Monitoring (SHM) of the road pavements refer to: (i) audio-visual inspections and image analysis [20, 23]; (ii) heavy and light instrumented vehicles, unmanned aerial vehicles, satellite, smartphones on vehicles using several devices (e.g., the smartphone's gyroscope) [24, 25]; (iii) ultrasonic guided waves, ultrasonic wave propagation, and ultrasonic tomography [26–28]; (iv) electromagnetic, e.g. the Ground Penetrating Radar (GPR), or using nuclear and non-nuclear electromagnetic gauges, or microwave imaging to scan the surface layers [29, 30]; (v) seismic methods (e.g., MASW), or devices such as the Light Weight Deflectometer (LWD), the Falling Weight Deflectometer (FWD), and the Rolling Wheel Deflectometer (RWD) [31, 32]; (vi) Ground Penetrating Radar (GPR) and thermo-cameras [33–35]; (vii) self-powered, or low- and ultralow-powered, wireless sensor networks [36–38].

1.3 FEM Models Applied on Road Pavements

This section contains relevant examples of the application of FEM on road pavements, which were used to simulate and study the behavior of these pavements under different load conditions, and to forecast their performances and sustainability over time. The examples mentioned below were grouped based on the software used to build the model, as follows: (a) ABAQUS was used to derive natural frequencies, dynamic moduli, fatigue cracking performance, and response to a given load. The above mentioned software was used also to simulate crack propagation, and interactions between deformable tires and pavements [39–41]; (b) ANSYS was used to study the bridge-vehicle interactions, and the resultant vibrations perceived by pedestrians [42]; (c) COMSOL Multiphysics was used to simulate structural responses to dynamic loads [43, 44]; (d) CAPA-3D was used to model tire-road contacts, and distress mechanisms [45]; (e) Matlab was used for the prediction of ground-borne vibrations due to road unevenness-tire interactions [46]; (f) the combination of SAFEM and ABAQUS was used to analyze mechanical performances of asphalt pavements (e.g., deflection, and strain/stress), and their dynamic responses to moving loads [47]; (g) SAFEM was used to simulate deterioration of asphalt pavements under heavy vehicle loads [48]; (h) SURFER was used to derivate ground borne vibrations, induced by vehicles, for different soils [49].

Despite the fact that, in the last decades, several solutions were proposed to improve the overall quality of the road pavements, it is still difficult to apply them in real contexts. Possible motivations can be the complexity, the cost, and the poor attention to the them by the authorities. For this reason, this paper aims at presenting a simple, and cheaper solutions that can be easily used to effectively improve the sustainability of the road pavements, in terms of smart monitoring and efficient maintenance. In more detail, the objectives of the study presented in this paper are: (1) to present a NDT-based monitoring system able to gather the vibro-acoustic responses of the road pavement to different loads; (2) to set up a FEM model able to simulate road pavement cracks and forecast any possible variations of the responses cited above due to a change of the structural status of the pavement.

Note that, this study is part of the research projects SICURVIA (project funded by Region Calabria, Italy).

2   Method

This paper describes a method that was developed to increase the sustainability of road pavements acting on two of the three approaches discussed in the previous section (see Fig. 1). In particular, the method is based on (1) the real time assessment of the structural conditions of the road (using an innovative monitoring system described in Sect. 2.1), and (2) on the prediction of the occurrence of road cracks due to repeated or extraordinary loads (using the FEM model described in Sect. 3). The method could be applied by authorities and companies that are responsible for the management of the road infrastructures, or could be used as a key component of the smart cities because of its ability to provide crucial information about the availability and reliability of the road infrastructures, which can be shared (Internet of Thing, IoT, approach) with users, or vehicles or other infrastructures (i.e., allow vehicle-to-infrastructure communication, and infrastructure-to-infrastructure communication).

2.1   NDT-Based Monitoring System and Data Gathering

An innovative NDT-based monitoring system was designed, based on the concept of vibro-acoustic signature. In more detail, the ground-borne response of the road pavement to vibration and noise generated by a mechanical source (e.g., vehicle, hammer, Light Weight Deflectometer, LWD) are considered as the vibro-acoustic signature of the road pavement. It is expected that the signature will be affected by any significant change of the structural conditions of the pavement. Hence, a system that is able to gather the above-mentioned signature can be used for monitoring purposes. The system used in this study consists of a broad-band and omnidirectional microphone located (NDT) on the surface of the road, which is able to gather the sound produced by the waves generated by a mechanical source that travels on the road. Importantly, this microphone must not be affected by the air-borne noise, and for this reason a sound-insulating dome was used to avoid undesirable effects due noise and wind. The structural condition of the road pavement was changed by means of drilled holes. Figure 2a shows the road pavement used in this study. Two different conditions were taken into account in this study, i.e., the structural condition 0, without holes, and the structural condition 1, with one line of 15 holes (diameter = 0.01 m, same depth as the one of the asphalt concrete layers = 15 cm, spaced 5 cm each other). A well-known mechanical source was used to load the road pavement, i.e., a LWD (see Fig. 2 left). This device is commonly used to measure the elastic modulus of pavements and is able to produce an impulse load that is suitable for the purposes of the study. The data set consists of 100 acoustic signals (50 for each condition), which were gathered with a sample frequency of 192 kHz.

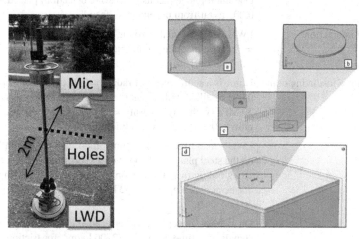

Fig. 2. (Left) experimental set up used to gather the data set used in the study, i.e. LWD, microphone + dome (Mic), and holes (structural condition 1); (Right) main elements of the FEM model: (a) isolating dome of the microphone; (b) LWD's steel plate; (c) dome-cracks-steel plate; (d) road pavement.

3 Set up of a FEM Model for Crack Prediction

The software COMSOL Multiphysics® was used to build the FEM model. In particular, the "Acoustic-Solid Interaction" interface, and the study "Pressure Acoustics Transient" were selected. Figure 1 right shows the main elements of the model, i.e. the road pavement, the drilled holes, the LWD base plate, and the dome of the probe. While, Table 1 summarizes the input parameters used to feed the FEM model.

Table 1. Input parameters of the FEM model.

Parameter	Details
Sampling	Sampling frequency of 2000 points per 200 ms, i.e., 10000 Hz, was selected
Geometry	LWD plate (height = 0.01 m, radius = 0.15 m). Pavement layers (thicknesses of the wearing course d_0 = 0.1 m, of the subbase course d_1 = 0.3 m, and of the subgrade course d_2 = 10 m; length of the layers = 10 m; width of the layer = 10 m). Boundary condition = "low reflection"
Sensors	One microphone (acoustic pressure detection) placed h_0 = 0.005 m far from pavement surface
LWD-road interaction	LWD steel base plate (Young's modulus = 200 GPa, density = 7850 kg/m^3; plate dimensions (diameter = 0.3 m, and height = 0.01 m)
Air-born noise insulating dome	Dome is a semi-sphere of radius = 0.05 m, thickness d_d = 0.005 m, filled with a material with Young's modulus = 50 MPa (with clay), density = 1800 kg/m^3, and Poisson's ratio = 0.3. Boundary conditions = "low-reflecting"
Impulse load	LWD's load function over time produced a maximum force on the steel plate cited above of 9 kN (that is due to a mass of 10 kg falling from 0.83 m, on five rubber buffers with an elastic constant of about 362 kN/m). The pulse time is 15 ÷ 30 ms
Layers	On average, lower moduli correspond to slower and bigger signals over time. Densities (2200 kg/m^3 for friction course, 2000 kg/m^3 for subbase course, and 1800 kg/m^3 for subgrade), and moduli (E_0 = 1000 MPa for friction course, E_1 = 500 kg/m^3 for subbase course, and E_2 = 90 kg/m^3 for subgrade course)
Temperatures	Bituminous layers elastic modulus is affected by the temperature, and it, in turn, affects the speed of the propagating waves. During the experiments, the road surface temperature was in the range 25–35 °C, and the reference temperature for air was assumed equal to 20.0 °C
Induced cracks	Drilled holes were modelled according to Sect. 3. Importantly, the air trapped into cracks was modelled itself
Speed of propagation	Vibrations propagate into the road pavement with a speed v_1 = 670 m/s, while the sound speed in air v_2 was derived from the air temperature

4   Results

The results of this study refer to the calibration of the FEM model presented in the previous section. In more detail, the calibration was carried out using meaningful features of the measured signals related to the peaks amplitude and time-lags. The peaks that were taken into account are shown in Fig. 3.

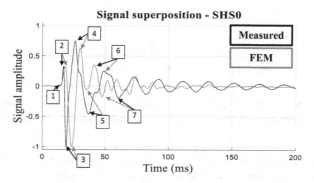

Fig. 3. Peaks used to represent the signals (features).

The time lags refer to the x-axis and the following feature were derived (the numbers point out the peak showed in Fig. 3): x2-x1, x3-x2, x4-x3, x5-x4, x6-x5, and x7-x6. The peak amplitude refers to the y-axis, and the features that were derived are: y1/y3, y2/y3, y3/y3, y4/y3, y5/y3, y6/y3, and y7/y3. After two rounds of calibration, the FEM model was able to provide the signals of Fig. 4 using the input parameter reported in Table 2.

Table 2. Main inputs of the calibrated FEM model.

ID	Main inputs											
	E_0	E_1	E_2	Ta	$Tpav$	d_d	$\emptyset_d$	v_1	v_2	μ	σ	F
	MPa	MPa	MPa	K	K	M	M	m/s	m/s	ms	ms	kN
1st	1000	500	90	293.15	293.15	0.01	0.05	670	n.a.	15	2	9
2nd	2000	600	200	299.75	304.25	0.008	0.075	1208	347.4	15	1.5	9

Symbols. ID = Round of calibration; Ei = moduli of the layers, with i = 0, 1, and 2; Ta = air temperature; $Tpav$ = pavement temperature; d_d = dome thickness; $\emptyset_d$ = dome diameter; $v_{1,2}$ = speeds of propagation in the transmission media 1 and 2; μ, σ = impulse peak position and width, respectively; F = impulse force.

Figure 4 shows the comparison between the measured vibro-acoustic signatures and those simulated using the FEM model. Based on the results shown in Fig. 3, it is possible to state that the procedure used to build and set up the FEM model can be effectively used to try to replicate real damage of a road pavement. Hence, this method can be also used to forecast the occurrence of any type of damages (e.g., surface or hidden cracks, or cracks due to thermal excursion, or fatigue failures).

Finally, the validation of the study was carried out comparing a set of the measured signals (42 signals) with a set of simulated data (42 signals). In particular, experiments showed an instrumental uncertainty (LWD) of about ± 5% (8.55–9.45 KN), which can

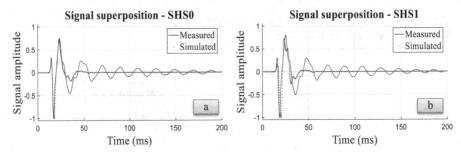

Fig. 4. Superposition of the measured and simulated vibro-acoustic signature of the road pavement in the structural conditions 0 (without holes) and 1 (with one line of holes).

be considered, together with the holes, the main cause of signal changes. Hence, further simulations were carried out, using the same parameters reported in Table 2, and changing the LWD load to obtain a simulated data set that takes into account the experimental uncertainties. Subsequently, a third round of calibration was carried out in order to find two calibration factors, i.e. one for the structural condition 0, and one for the structural condition 1. In more detail, each simulated signal (referred to the given condition) was divided for a reference signal (properly selected, measured signal) obtaining one corresponding matrix of factors. Then, from the elements of the given column an average was derived. The resulting two vectors (one *per* conditions) were used as calibration factors. These latter were multiplied by the simulated signals obtaining the final simulated data set. Finally, the two abovementioned data sets (measured and simulated after three rounds of calibrations) were used as input of a hierarchical clustering algorithm (implemented in Matlab®) that aimed at classifying the signals into two classes, i.e., structural condition 0 and 1. The result of the classification was expressed in terms of model accuracy (i.e., the ratio between the number of signals correctly classified and the total number of signals belonging to the testing data set to be classified), derived from confusion matrixes (i.e., a matrix that shows how many observations were associated to each cluster). The average model accuracy resulted of about 85% (i.e., 83% and 86% for the structural condition 0 and 1, respectively), which can be considered sufficient to validate the FEM model.

5 Conclusions

Smart cities need roads with high levels of sustainability, and this goal can be reached using, e.g., smart materials, Non-Destructive Test (NDT)-based monitoring systems, and Finite Element Method (FEM)-based damage prediction models. Consequently, the main objectives of this study are (i) providing an overview of the current approaches, and (ii) presenting a NDT-, and FEM-based monitoring method (road damage detection and prediction). The proposed method has the potentialities to contribute to a more sustainable environment because it is based on (1) the real time assessment of the structural conditions of the road (using an innovative monitoring system), and (2) on the prediction of the occurrence of road cracks due to repeated or extraordinary loads (using the FEM model). The FEM model was calibrated using a data set gathered during an experimental investigation on road pavements that was progressively damaged with drilled holes. A simple NDT apparatus was used to gather the vibro-acoustic signatures of road pavement (data set). Results show that, the procedure used to build and set up the FEM model can be effectively used to replicate real damage of a road pavement, and that this method can be used to forecast the occurrence of any type of damages. Hence, the proposed solution can positively affect both the maintenance efficiency of authorities and companies that are responsible for the management of the road infrastructures, and the road sustainability by means of the ability to share crucial information (road availability and reliability) with users, or vehicles or other infrastructures (IoT approach).

Finally, based on the promising results, possible further research steps are: (1) reproducing typical cracks of the road pavements and, then, using the monitoring system to detect the vibro-acoustic responses and the FEM model to predict the variation of the above mentioned responses due to the propagation of the induced cracks; (2) using the FEM model to design new smart materials and forecast their behavior; (3) improving the system using low-power electronics (e.g., MEMS sensors, low-power and ultra-low power wireless transmitters) for large-scale production and application. The future researches listed above will require interactions with other research fields such as the machine learning one to face the increase of the data set size due to the use of the monitoring system (big data), and the micro-electronics to improve the performances of the monitoring system.

References

1. Neelakanta, P.: Smart Materials (2013). https://doi.org/10.1201/9781420049763.ch58
2. Praticò, F.G., Vaiana, R., Giunta, M.: Pavement sustainability: permeable wearing courses by recycling porous European mixes. J. Archi. Eng. (2013). https://doi.org/10.1061/(ASCE)AE. 1943-5568.0000127
3. Wang, T., Xiao, F., Zhu, X., Huang, B., Wang, J., Amirkhanian, S.: Energy consumption and environmental impact of rubberized asphalt pavement. J. Cleaner Prod. (2018). https://doi.org/10.1016/j.jclepro.2018.01.086
4. Praticò, F.G., Moro, A., Noto, S., Colicchio, G.: Three-year investigation on hot and cold mixes with rubber. In: 8th International Conference on Maintenance and Rehabilitation of Pavements, MAIREPAV (2016). https://doi.org/10.3850/978-981-11-0449-7-085-cd
5. Le, J.L., Marasteanu, M.O., Turos, M.: Mechanical and compaction properties of graphite nanoplatelet-modified asphalt binders and mixtures. Road Mater. Pavement Des. (2019). https://doi.org/10.1080/14680629.2019.1567376
6. Du, Y., Chen, J., Han, Z., Liu, W.: A review on solutions for improving rutting resistance of asphalt pavement and test methods. Constr. Build. Mater. (2018). https://doi.org/10.1016/j.conbuildmat.2018.02.151
7. Wang, H., Yang, J., Liao, H., Chen, X.: Electrical and mechanical properties of asphalt concrete containing conductive fibers and fillers. Constr. Build. Mater. (2016). https://doi.org/10.1016/j.conbuildmat.2016.06.063
8. Li, H., et al.: Investigation of the effect of induction heating on asphalt binder aging in steel fibers modified asphalt concrete. Materials (2019). https://doi.org/10.3390/ma12071067
9. Sun, D., Li, B., Ye, F., Zhu, X., Lu, T., Tian, Y.: Fatigue behavior of microcapsule-induced self-healing asphalt concrete. J. Cleaner Prod. (2018). https://doi.org/10.1016/j.jclepro.2018.03.281

10. Sun, Y., Wu, S., Liu, Q., Li, B., Fang, H., Ye, Q.: The healing properties of asphalt mixtures suffered moisture damage. Constr. Build. Mater. (2016). https://doi.org/10.1016/j.conbuildmat.2016.10.048

11. García, A., Norambuena-Contreras, J., Bueno, M., Partl, M.N.: Single and multiple healing of porous and dense asphalt concrete. J. Intell. Mater. Syst. Struct. (2015). https://doi.org/10.1177/1045389X14529029

12. Hasheminejad, N., et al.: Digital image correlation to investigate crack propagation and healing of asphalt concrete. In: Proceedings of the Eighteenth International Conference of Experimental Mechanics, p. 5381. MDPI, Basel (2018). https://doi.org/10.3390/icem18-05381

13. Patti, F., Mansour, K., Pannirselvam, M., Giustozzi, F.: Mining materials to generate magnetically-triggered induction healing of bitumen on smart road pavements. Constr. Build. Mater. (2018). https://doi.org/10.1016/j.conbuildmat.2018.03.160

14. Zhang, H., Bai, Y., Cheng, F.: Rheological and self-healing properties of asphalt binder containing microcapsules. Constr. Build. Mater. (2018). https://doi.org/10.1016/j.conbuildmat.2018.07.172

15. Fedele, R., Merenda, M., Praticò, F.G., Carotenuto, R., Della Corte, F.G.: Energy harvesting for IoT road monitoring systems. Instrum. Mesure Metrologie **17**(4), 605–623 (2018). https://doi.org/10.3166/I2M.17.605-623

16. Yoomak, S., Ngaopitakkul, A.: Optimisation of lighting quality and energy efficiency of LED luminaires in roadway lighting systems on different road surfaces. Sustain. Cities Soc. (2018). https://doi.org/10.1016/j.scs.2018.01.005

17. Praticò, F., Della, F., Merenda, M.: Self-powered sensors for road pavements. In: 4th Chinese–European Workshop on Functional Pavement Design, CEW (2016). https://doi.org/10.1201/9781315643274-151

18. Hasni, H., Alavi, A.H., Chatti, K., Lajnef, N.: A self-powered surface sensing approach for detection of bottom-up cracking in asphalt concrete pavements: theoretical/numerical modeling. Constr. Build. Mater. (2017). https://doi.org/10.1016/j.conbuildmat.2017.03.197

19. Bansal, R.C.: Electric vehicles. In: Emadi, A. (ed.) Handbook of Automotive Power Electronics and Motor Drives, 1st edn. Taylor & Francis, Boca Raton (2017)

20. Miller, J.M.: Hybrid electric vehicles. In: Emadi, A. (ed.) Handbook of Automotive Power Electronics and Motor Drives, 1st edn. Taylor & Francis, Boca Raton (2017). https://doi.org/10.1201/9781420028157

21. Hola, J., Schabowicz, K.: State-of-the-art non-destructive methods for diagnostic testing of building structures – anticipated development trends. Arch. Civ. Mech. Eng. (2010). https://doi.org/10.1016/S1644-9665(12)60133-2

22. Praticò, F.G., Vaiana, R.: A study on the relationship between mean texture depth and mean profile depth of asphalt pavements. Constr. Build. Mater. (2015). https://doi.org/10.1016/j.conbuildmat.2015.10.021

23. Zhang, Y., et al.: Kinect-based approach for 3D pavement surface reconstruction and cracking recognition. IEEE Trans. Intell. Transp. Syst. **19**, 3935–3946 (2018)

24. Carlos, M.R., Aragon, M.E., Gonzalez, L.C., Escalante, H.J., Martinez, F.: Evaluation of detection approaches for road anomalies based on accelerometer readings-addressing who's who. IEEE Trans. Intell. Transp. Syst. (2018). https://doi.org/10.1109/TITS.2017.2773084

25. Cafiso, S., D'Agostino, C., Delfino, E., Montella, A.: From manual to automatic pavement distress detection and classification. In: 5th IEEE International Conference on Models and Technologies for Intelligent Transportation Systems (MT-ITS) (2017). https://doi.org/10.1109/mtits.2017.8005711

26. Pahlavan, L., Mota, M.M., Blacquière, G.: Influence of asphalt on fatigue crack monitoring in steel bridge decks using guided waves. Constr. Build. Mater. (2016). https://doi.org/10.1016/j.conbuildmat.2016.05.138

27. Mounier, D., Di Benedetto, H., Sauzéat, C.: Determination of bituminous mixtures linear properties using ultrasonic wave propagation. Constr. Build. Mater. (2012). https://doi.org/10.1016/j.conbuildmat.2012.04.136

28. Hoegh, K., Khazanovich, L., Yu, H.: Ultrasonic tomography for evaluation of concrete pavements. Transp. Res. Rec. J. Transp. Res. Board (2011). https://doi.org/10.3141/2232-09

29. Praticò, F.G., Moro, A., Ammendola, R.: Factors affecting variance and bias of non-nuclear density gauges for PEM and DGFC. Baltic J. Road Bridge Eng. **4**, 99–107 (2009)

30. Bevacqua, M.T., Isernia, T.: Boundary indicator for aspect limited sensing of hidden dielectric objects. IEEE Geosci. Remote Sens. Lett. (2018). https://doi.org/10.1109/LGRS.2018.2813087

31. Iodice, M., Muggleton, J., Rustighi, E.: The detection of vertical cracks in asphalt using seismic surface wave methods. J. Phys. Conf. Ser. (2016). https://doi.org/10.1088/1742-6596/744/1/012059

32. Pitoňák, M., Filipovsky, J.: GPR application - non-destructive technology for verification of thicknesses of newly paved roads in Slovakia. Procedia Eng. **153**, 537–549 (2016). https://doi.org/10.1016/j.proeng.2016.08.184

33. Solla, M., Lagüela, S., González-Jorge, H., Arias, P.: Approach to identify cracking in asphalt pavement using GPR and infrared thermographic methods: preliminary findings. NDT Int. (2014). https://doi.org/10.1016/j.ndteint.2013.11.006

34. Ouma, Y.O., Hahn, M.: Wavelet-morphology based detection of incipient linear cracks in asphalt pavements from RGB camera imagery and classification using circular Radon transform. Adv. Eng. Inform. (2016). https://doi.org/10.1016/j.aei.2016.06.003

35. Grace, R.: Sensors to support the IoT for infrastructure monitoring: technology and applications for smart transport/smart buildings. In: MEPTEC IoT Conference, San Jose, CA (2015)

36. Fedele, R., Della Corte, F.G., Carotenuto, R., Praticò, F.G.: Sensing road pavement health status through acoustic signals analysis. In: 13th Conference on Ph.D. Research in Microelectronics and Electronics (PRIME) (2017). https://doi.org/10.1109/prime.2017.7974133

37. Lenglet, C., Blanc, J., Dubroca, S.: Smart road that warns its network manager when it begins cracking. IET Intell. Transp. Syst. **11**, 152–157 (2017). https://doi.org/10.1049/iet-its.2016.0044

38. Xu, X., Cao, D., Yang, H., He, M.: Application of piezoelectric transducer in energy harvesting in pavement. Int. J. Pavement Res. Technol. (2018). https://doi.org/10.1016/j.ijprt.2017.09.011

39. Alavi, A.H., Hasni, H., Lajnef, N., Chatti, K.: Continuous health monitoring of pavement systems using smart sensing technology. Constr. Build. Mater. (2016). https://doi.org/10.1016/j.conbuildmat.2016.03.128

40. Hernandez, J.A., Al-Qadi, I.L.: Tire–pavement interaction modelling: hyperelastic tire and elastic pavement. Road Mater. Pavement Des. (2017). https://doi.org/10.1080/14680629.2016.1206485

41. Moghimi, H., Ronagh, H.R.: Development of a numerical model for bridge-vehicle interaction and human response to traffic-induced vibration. Eng. Struct. (2008). https://doi.org/10.1016/j.engstruct.2008.06.015

42. Pedersen, L.: Viscoelastic Modelling of Road Deflections for use with the Traffic Speed Deflectometer. Kgs. Lyngby: Technical University of Denmark (DTU). IMM-PHD-2013, No. 310 (2013)

43. Stamp, D.H., Mooney, M.A.: Influence of lightweight deflectometer characteristics on deflection measurement. Geotech. Test. J. (2013). https://doi.org/10.1520/GTJ20120034

44. Casey, D.B., Airey, G.D., Grenfell, J.R.: Relative pavement performance for dual and wide-based tire assemblies using a finite element method. J. Test. Eval. **45**, 1896–1903 (2017). https://doi.org/10.1520/JTE20160589

45. Lak, M.A., Degrande, G., Lombaert, G.: The effect of road unevenness on the dynamic vehicle response and ground-borne vibrations due to road traffic. Soil Dyn. Earthq. Eng. (2011). https://doi.org/10.1016/j.soildyn.2011.04.009

46. Liu, P., Wang, D., Oeser, M.: Application of semi-analytical finite element method coupled with infinite element for analysis of asphalt pavement structural response. J. Traffic Transp. Eng. (2015). https://doi.org/10.1016/j.jtte.2015.01.005

47. Liu, P., Wang, D., Hu, J., Oeser, M.: SAFEM – software with graphical user interface for fast and accurate finite element analysis of asphalt pavements. J. Test. Eval. **45**, 1301–1315 (2017). https://doi.org/10.1520/JTE20150456. ISSN:0090-3973

48. Liu, P., Otto, F., Wang, D., Oeser, M., Balck, H.: Measurement and evaluation on deterioration of asphalt pavements by geophones. J. Int. Measure. Confederation (2017). https://doi.org/10.1016/j.measurement.2017.05.066

49. Astrauskas, T., Grubliauskas, R.: Modelling of Ground Borne Vibration Induced by Road Transport. Vilnius Gediminas Technical University, Vilnius (2017). https://doi.org/10.3846/mla.2017.1060

Quantifying the Carbon Dioxide Emissions Resulting from Awareness-Raising Actions of Sustainable Mobility

Pétilin Souza(✉) , Filipa Paiva , Lígia Silva , and Paulo Pereira

University of Minho, Rua da Universidade, 4710-057 Braga, Portugal
petilindesouza@gmail.com, filipa_paiva@sapo.pt,
{lsilva,ppereira}@civil.uminho.pt

Abstract. Currently, the transport sector is a major contributor to global greenhouse gases emissions. The evaluation of projects that promote the decarbonization in cities is mainly focused on calculating the reduction of CO_2 emissions. The calculation methodology takes into account the characteristics of the city car fleet where these projects are implemented. In the case of the "SchoolBus" project, promoted by the Braga Municipal Council during the 2018–2019 school year, the average value of reduction was of 34.9 tonCO_2eq/year. The number of vehicles circulating in the study area, the emission factor related to the type of vehicle and its fuel and the distance traveled by them were the main parameters used to calculate CO_2 emissions. The impacts of the "SchoolBus" project on stakeholders and awareness-raising actions of sustainable mobility can be quantified in terms of CO_2, by analyzing the population's interest in joining more sustainable modes of transport. Thus, it is possible to evaluate the potential of the awareness-raising actions, applied in the study area, in the decarbonization in a larger area (i.e. the whole municipality). At the same time, it is possible to compare and analyze the calculated values in relation to the quantified potential range of effects of policy measures on CO_2 emissions using transport demand for the year 2030, indicated by the European Commission and adapted to the area of study, by introducing measures of sustainable mobility in the territory.

Keywords: Cities decarbonization · Sustainable urban mobility · Public participation

1 Introduction

Nowadays, people are moving from rural areas to cities in order to search better opportunities of labor. According to the study carried out by the United Nations, an urban population growth was observed in Portugal in 1995 [1]. Worldwide, it was determined that 55.7% of the population lives in urban areas, with an expected increase to 68.4% by 2050 [1]. However, in Portugal it is expected a future increase of 79.3% of the urban population [1].

© ICST Institute for Computer Sciences, Social Informatics and Telecommunications Engineering 2020
Published by Springer Nature Switzerland AG 2020. All Rights Reserved
P. Pereira et al. (Eds.): SC4Life 2019, LNICST 318, pp. 15–30, 2020.
https://doi.org/10.1007/978-3-030-45293-3_2

The growth of the urban population has created impacts on the environment, whether directly or indirectly. According to IPCC [2], the direct and indirect impacts are as follows: (i) the increase in temperatures in cities; (ii) the higher concentration of air pollutants, including carbon dioxide emissions; (iii) the increase in noise, mainly from road traffic; and (iv) floods or droughts, creating water crises in urban areas [2].

Road traffic is one of the main sources of pollution, both air and noise, causing adverse effects on human health [3]. Increasingly, the combustion of fuels from road vehicles (60% of air pollutants emitted in urban and surrounding areas) is one of the largest air pollutants [4].

Greenhouse gases (GHG), when present in the atmosphere, absorb and emit part of the infrared radiation, making it impossible the Earth to lose heat to space. Carbon dioxide (CO_2) is a greenhouse gas, although it is not a toxic gas, and its emissions are the main cause of global warming. The increase in CO_2 emissions is therefore one of the most important issues in an urban area [2, 4].

The case study of this article is based on the project developed by Braga City Council (CMB), Braga Urban Innovation Laboratory Demonstrator (BUILD), with the aim of reducing greenhouse gas emissions. The BUILD project aims to solve car traffic issues, improving the urban environment, based on the implementation of technological and innovative solutions in the city. The project also aims to raise awareness in the community, in order to make them aware of the importance of decarbonization and the consequent impact on raising quality of life. The area chosen for the implementation of the BUILD project is located in the parish of São Vicente, and there are seven educational institutions in this area.

Based on the analysis of the results of the actions carried out under the BUILD project, this study aims to quantify the CO_2 emissions resulting from sustainable mobility awareness actions, analyzing and comparing the quantified values in relation to the potential CO_2 emission reduction values established by the European Commission, which have been adapted to the area under study.

Currently, there are a great range of technological equipment capable of providing effective tools, such as electronic sensors to collect information on traffic and air pollution concentrations, thus enabling support for public policy decisions. However, a change in community habits is highly important in order to decarbonize the city and contribute to the reduction of environmental impacts. This change is only possible with the application of sustainable policies and through the participation of the community in this objective.

The present study evaluates different approaches of urban planning in the context of the BUILD project, in order to improve urban quality of life and mitigate the climate changes. More specifically, BUILD aims to promote sustainable and inclusive urban mobility, prioritizing pedestrian mode, cycling and public transport over individual transport, with the decarbonization of the study area as its main purpose. Moreover, the BUILD stimulates the potential of awareness actions, as well as the application of measures developed within the framework of Sustainable Mobility to reduce emissions in the study area. The influence of different approaches from BUILD project on CO_2 emissions in the study area were compared and evaluated according to the method indicated by the European Commission, which was adapted to the study area, defined as PER Framework.

In the following sections the theoretical approach, the adopted methods, the case study exposition and the conclusions obtained from it are presented.

1.1 Urban Mobility and Traffic Pollution

Across the EU cars remain the dominant mode of passenger transport and account, approximately, 12% of total carbon dioxide (CO_2) emissions, the main greenhouse gas (GHG) [5]. Economic and population growth have been the most important factors in increasing CO_2 emissions from fossil fuel combustion [6]. In 2010, the transport sector accounted for 14% of total direct anthropogenic greenhouse gas emissions (gigatons of CO_2-equivalent per year, $GtCO_2$-eq/year) [2].

The GHG emissions from transport have not reduced in line with other economic sectors, although CO_2 emissions of new passenger cars have been steadily decreasing since 2007, reaching a value of 118.5 g/km in 2017 [5]. Thus, CO_2 emissions per km decreased by 10.4% between 2012 and 2017 [5]. To decarbonize cities, increasing their resilience to climate change, it is necessary to change lifestyles and current mobility patterns, which focus on individual transport [2]. Thus, the impact on environment by choosing between more or less sustainable transportation modes represents a challenge for decarbonization.

The planning process of transportation has changed throughout time, but to improve the cities is required a careful planning [7]. Traditional urban planning was carried out by traffic engineers, focusing on traffic modes, infrastructure and transportation modals [8]. Currently, urban mobility planning is focused on people, presenting integrated actions to achieve cost-effective solutions in a balanced development of all relevant transport modes, regarding the transition towards cleaner and more sustainable transport modes. On the other hand, city planning began to be carried out by interdisciplinary teams with stakeholder involvement.

The decarbonization of cities has been materialized by several policy objectives focused on urban mobility, such as the sharp reduction in CO_2 emissions from the transport sector to zero between 2040 and 2050; a marked reduction in emissions of air pollutants; almost zero fatalities in road transport in 20 to 40 years, depending on the current situation; drastic reduction of traffic congestion; improving the quality of urban life in general [7]. This shows the need to continually act on mobility in order to achieve a more sustainable scenario.

1.2 Sustainable and Inclusive Urban Mobility as a Goal

The year of 2015 marked a turning point towards sustainable development around the world. World leaders adopted a new global framework for sustainable development at the 70th UN General Assembly on 25 September 2015: the 2030 Agenda for Sustainable Development, based on the Sustainable Development Goals (SDGs). The 2030 Agenda represents a commitment to eradicate poverty and achieve sustainable development by 2030. The 17 SDGs and their 169 associated objectives are global in nature, universally applicable and interlinked [6].

The Paris Climate Agreement (COP21), the Addis Ababa Action Agenda, as an integral part of the 2030 Agenda, and the Sendai Framework for Disaster Risk Reduction also began to be implemented in 2015 [6]. The 2015 Paris Agreement represents a remarkable landmark in the global fight against climate change. In turn, the SDG 13 represents the EU's path to a low-carbon and climate resilient economy, through setting an ambitious economy-wide domestic target of reducing GHG emissions in at least 40% by 2030. The target is based on global projections that are in line with the medium term ambition of the Paris Agreement [6].

The SDG 11 marks, in the following, the importance of the sustainable mobility, providing access to safe, affordable, accessible and sustainable transport systems for all. It also enhances inclusive and sustainable urbanization and capacity for participatory, integrated and sustainable human settlement planning and management in all countries.

Urban mobility planning is a challenging and complex task. Planners need to manage many demands at the local and international levels, such as the European climate change and energy efficiency targets. The complexity increases in the event of political change and severe financial constraints, as is the situation in many European countries today. A Sustainable Urban Mobility Plan (SUMP) contributes to achieving the European energy and climate objectives set by EU leaders [8].

A SUMP is a strategic plan aimed at improving the quality of life of cities and surrounding areas, taking into account the mobility needs of people and businesses, which is based on existing planning practices and considers the principles of integration, participation and evaluation. The basic principles of a Sustainable Urban Mobility Plan are: Long-term vision and clear implementation plan; Participatory approach; Balanced and integrated development of all modes of transport; Horizontal and vertical integration; Evaluation of current and future performance; Regular monitoring, review and reporting; and consideration of external costs for all modes of transport [8].

1.3 Mobility Projects to Promote Decarbonization

Echeverri [9] and Grandim [10] refer that more than 70% of carbon dioxide emissions are produced by cities, making the discussion of urban challenges relevant to achieve a reduction in them. These urban challenges come from different dimensions, including urban infrastructure, transport, buildings and waste, presenting at the same time difficulties and opportunities at the technical and economic level [10]. Therefore, these mitigation challenges address the following levels: (i) institutions; (ii) behaviors; (iii) values; and (iv) technologies.

A relational perspective of these levels of urban dynamics allows us to understand that the various types of relationships that make up cities are what really create and change cities. Looking at how urban development is structured by its political economy and investigating how material forces and non-human agency are molding urban life, an adaptable relationship between both everyday practices and urban infra-structure becomes evident and results in unequal patterns of urban energy request and well-being [10].

With the implementation of new energy and mobility infrastructures in urban areas, there is potential for reducing carbon emissions [10]. Grandim et al. [10] address the complexity behind the implementation of new technologies or the investment of a public transport project but, more than that, a cultural and political social challenge.

The European Commission's report determined the expected impact of urban sustainable mobility measures on reducing CO_2 emissions for urban areas and examined the NUT-3 areas for Europe in order to establish the interval of the potential impact [11]. The report sets out the following 21 sustainable mobility measures, each with 7 categories: (i) public transport services; (ii) urban logistics and distribution; (iii) mobility management; (iv) incorporation of transport modes; (v) road transport; (vi) marketing and educational campaigns; (vii) access restriction schemes. In order to quantify the defined measures, a range corresponding to the total at European level was established and, thus, the total value of the contribution of each dimension was defined for each country [11]. These dimensions address the need to replace unsustainable transport with sustainable transport. Taking this into account, the study conducted by Souza [12] defined the values adopted for the country where the area under analysis is located, and stipulated the reduction coming from the expected impact values for the dimensions of the study area [12].

In recent decades, the number of parents travelling by individual transport to take their children to school has increased substantially, thus reducing the number of journeys by students on public transport. The aim of school mobility plans is to improve accessibility to schools by enabling pupils to travel independently to schools in a sustainable and safe manner [11]. There are several alternative modes of sustainable mobility, including collective bus travel, walking, cycling or car sharing between parents, which promote more sustainable behaviour in children with regard to environmental and safety issues [11].

In 2014, in Cordoba, Spain, a 14-week free Walking School Bus pilot project was developed. This project, carried out in a public school with the participation of 450 students, and consisted in students going to school, accompanied by monitors. The monitors had access to an application that allowed them to add the children's participation data and provide the group's location.

School Mobility is an important aspect for the control of road traffic and, as well as for the mobility improvement in the city of Braga [13]. Thus, as part of the European Mobility Week, in 2017, the CMB developed activities in the city in order to experiment solutions to reduce road traffic in the BUILD area. The activities developed were the "PeddyBus" and the "SchoolBus". The "PeddyBus" activity consisted in the movement of students, in groups and under the supervision of adult volunteers, to schools. In a first trial, about 10 children participated in the activity [13]. At the same time, the pilot project "SchoolBus" was under development, and its function was the free transportation of students by bus to previously established schools, with the participation of 700 children in the project [13].

2   Methods

Several measures that change urban infrastructure or mobility patterns are good options to reduce CO_2 emissions, for instance central urban areas closure for vehicles with pedestrian networks (SMM 5); integrated systems of public transportation (SMM 3); implementation of complete streets (SMM 14); scholar mobility plans (SMM 8) and implementation of low emission zones (SMM 21).

As previously mentioned, although traditional planning has a great focus on traffic, the discussion of sustainable urban mobility planning considers people as more important [8]. Therefore, it is extremely necessary to have an approach with participatory actions through a transparent relation. And so, the population and general stakeholders participate since the beginning of the plan development until the implementation phase. This participatory approach guarantee that the population own a Sustainable Urban Mobility Plan and the policies related to it [8].

The acceptance of a sustainable mobility plans, through long-term actions, and their permanence on the territory have greater chances to last with the involvement of the population [11, 15]. Therefore, willing to achieve the objectives previous presented, the actions will be developed along with stakeholders to promote sustainable and inclusive urban mobility.

2.1 Awareness-Raising Actions

In order to promote the "SchoolBus" project among the scholar community, several activities aimed at the target public, were applied. The awareness-activities developed were: (i) mobility as a game; and (ii) sustainable mobility workshops. The barriers of the involvement of the population in planning are the idea of a high need of effort and time [15]. Regarding it, it is important to use approaches, such as playful participatory processes (PPP), that insure a more attractive activities and beaten these mentioned issues. A few examples of PPP are story-telling, walking activities, sketching or drawing and games [15]. Thus, regarding the awareness-raising actions, more specifically for the engagement of the students, the activity will consist in two games that address the theme of sustainable mobility, in order to disseminate this concept and, therefore, increase the perception of different modes of transport's impact on the environment.

The Sustainable Mobility Workshop (SMW) developed to reach the goals of raising awareness engaged general stakeholders, focusing on students, willing to help to transition from the ongoing mobility paradigm towards more sustainable options,, and was based on the activity developed in Sao Paulo, Brazil, by ITDP Brasil [14]. Therefore, the students were first invited to fill a board with their daily transportation mode on the commuting to school, between the options of by foot, by bicycle by motorcycle, by car, by taxi or by public transport.

To motivate the participation it is possible to apply free-style drawing, beyond the options of playing a game or drawing, since it is possible to comprehend either objects and situations in the geographic environment [15].

Regarding it, the students later should draw their route in a mental map form, considering the path from home to school. It was important to highlight a few reference points, as well as some obvious figures for them, like people, things or landscapes, always emphasizing their transportation mode.

In an attempt to gather more individual information, the students could share theirs drawings, by explaining to their friends their creations. At the same time, the technical team would support the description by asking some previous agreed questions, about their feeling during the journey, their routine activities meanwhile travelling, if it was pleasant for them or would they consider some aspect to change. If the students mentioned an aspect to change, they could mark in red these modifications.

To assess the impact of the actions it will be calculated the number of people who confirmed to be willing to transition towards a more sustainable transportation mode, and this quantity of people will be considered as vehicles that will no longer drive in that area, allowing the calculation of the CO_2 emission reduction.

2.2 PER Framework

The Joint Research Centre (JRC) from the European Commission [11] calculated in their report 21 sustainable mobility measures' potential impact on reducing CO_2 emissions in an urban environment, thorough defining the effects of those measures for each NUTS-3 zone in Europe. To better understand the NUTS (Nomenclature of Territorial Units for Statistics) zoning, it is a classification through a hierarchical system which separates the European Union (EU) economic territory regarding its dimensions considering different intentions (European Commission, n.d.). The division of the territory comes in basically three different levels, the first is for major socio-economic regions, the second basic regions in which the regional policies are applied, finally in a third level for small regions that need specific analysis to more accurate diagnosis. In the case of Portugal, the division of NUTS-1 is Continent, Azores, and Madeira; NUTS-2 is five regions and two autonomous regions; finally, NUTS-3, concern groups of municipalities.

In the study of JRC [11] it was considered the demand of transportation by the year of 2030 through MODEL-T JRC, and the estimated CO_2 emissions resultant for each NUTS-3 zone, in order to quantify the potential interval of policies' effects on reducing CO_2 emissions. The potential effects of the policies of sustainable mobility were asset by an analysis which considered the differences between European territories and weighted scores from a group of experts regarding tendencies of transport behavior in each NUTS-3 zones. Once each European city distinguished itself for its size, density, population, and other aspects, the weighing method of scoring considered it, since it could certainly differ between cities [11]. Consequently it were analyzed different aspects of the cities, such as density, accessibility, employment, population and commuting characteristics of each particular zone.

The values presented by JRC [11] were adapted in the PER Framework (Table 1) to the dimensions of the study area.

To develop this scale-down approach, Souza [12] calculated each sustainable measure contribution regarding the values range of A-S-I dimensions for Portugal potential values. As a result, Portugal represented 1.27% of the European total emission of CO_2.

Afterwards, Souza [12] considered the more refined spatial dimension of sub-NUTS-3, that are more precise dimensions, and through the correlation of these sub zones with the Functional Urban Area (FUA) of each NUTS-3 zone it was possible to scale the values from Portugal to the dimensions of the city of Braga. The FUA analyzed in combination with NUTS-3 zones allowed a greater precision in the estimation of each transport activity specification by area. Through the method of Random Forest model, it was set a matrix of 642 (FUAs) × 21 (measures) × 3 (coefficients: A-S-I), that represented the net impact on emissions reduction across all modes in relative terms (as a share of estimated total urban transport emissions of each FUA) [16].

Some simplifications on the calculation were considered by Souza [12] since in the matter of the present study is analyzed a specific area of a city instead a comparison between different cities. Conclusively it was create a framework (PER Framework)

Table 1. PER Framework [12]

PER Framework	[tonCO$_2$eq/year]	
Sustainable mobility measures	Min	Max
Public transport services		
1 Investment and maintenance, including safety, security and accessibility	34	43
2 Public transport coverage & frequencies	44	55
3 Interoperable ticketing and payment systems	23	28
4 Taxi services (individual and collective)	28	35
5 Dedicated walking and cycling infrastructure investment and maintenance & Bike sharing schemes	38	47
City logistics and distribution		
6 Improvement of the efficiency of city logistics by the use of ICT	46	57
7 Measures to improve the energy efficiency and environmental performance of vehicles and/or use of alternative modes	30	37
Mobility management		
8 Corporate, school and personalized mobility plans (or workplace travel plans)	33	41
9 Car sharing & carpooling schemes	21	27
10 Telecommunications	49	61
Integration of transport modes		
11 Multimodal connection platforms	15	18
12 Multimodal travel information provision	41	51
13 Park and Ride areas	25	31
Road transport		
14 Reallocation of road space to other modes of transport, e.g. dedicated bus lanes	48	59
15 Parking management	38	47
16 Dynamic traffic management measures	20	25
17 Low speed zones	23	29
Marketing campaigns and education		
18 Information and marketing campaigns	30	38
19 Promotion of eco-driving	7	9
Access restriction schemes		
20 Congestion charging zones	72	90
21 Low emission zones	41	51

which allows the assessment and analysis of the potential reduction of CO$_2$ emission by each measure considered in the calculation.

The impact of the sustainable mobility measures can be asset through the PER once it calculates the quantity of CO$_2$ that will no longer be emitted, considering the dimension of the study area in comparison with the interval set by the framework. The calculation evaluation is presented in the following section.

3 Case Study: Awareness Activities Under Sustainable Mobility Theme - The BUILD Project

Braga City Council developed a project in the form of Living Lab, called Braga Urban Innovation Laboratory Demonstrator (BUILD), which focused on the decarbonization of the city, allowing an urban environment accessible to innovation, through the actions of local authorities, companies, the University of Minho (UM), the International Iberian Nanotechnology Laboratory (INL), the Center for Computer Graphics (CCG), citizens and local communities. As part of this study and after the success of the "SchoolBus" pilot project carried out during the mobility week, the CMB permanently implemented the "SchoolBus" project during the 2018/2019 school year. This project consists in transporting students by bus to the respective schools, located in the city centre and belonging to the BUILD intervention area, from points located on the border of the city centre, previously defined in the project. BUILD's intervention area is located in the city of Braga, more specifically in a parish in the northeast area, São Vicente, which integrates the historical centre area and the Braga East area. The area represents one of the most recent growth areas of the city's urban area, with around 14 hectares, being highly influenced by its proximity to the city centre [14]. The north and northwest border of the area is delimited by the Regiment Street of Infantry 8 and along the east border is the cemetery of Braga and the street Dom António Bento Martins Júnior. The south border of the area is delimited by the Monte de Arcos square, the Conselheiro Bento Miguel street, the Conselheiro São Januário street and Dr. Domingos Soares street. According to the Geographic Information Reference Database (BGRI) and analyzing the Census 2011, there are 75 buildings in the BUILD area, 62 of which are exclusively residential and the rest with residential and commercial use. The study area presents a high difference between the areas close to the historical center, where the residential buildings are new and large, and the northern areas, where the buildings are smaller, usually single-family, and older [14]. The area under analysis has a high road traffic, due to the high concentration of educational institutions. The area contains two leisure areas, among them there is the Monte d'Arcos Square, which acts as living area, and the public park called the Pachancho Children's Park (Fig. 1).

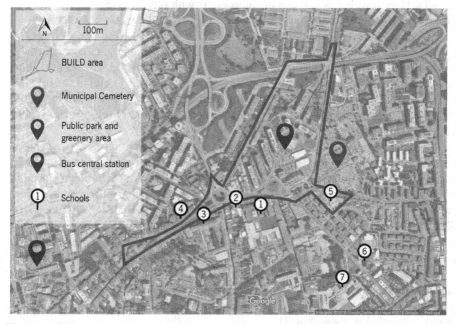

Fig. 1. BUILD area and reference points in which (1) Dom Diogo de Sousa School (2) Leonardo Da Vinci School (3) Sá de Miranda School (4) British Institute (5) Enguardas School (6) Teresiano School (7) Francisco Sanches School [12]

The study area presents considerable road traffic congestion, mainly during the period of entry and exit of students from the various educational institutions. It also presents a problem of lack of inclusive mobility and soft modes, under the existing conditions of urban public life and access, especially in relation to abandoned and damaged public spaces. Thus, the main objective of the "SchoolBus" project is to improve accessibility, safety and, above all, the mobility in the vicinity of schools, with a perspective of decarbonizing the BUILD area. The "SchoolBus" project started in September 2018 and, in January 2019, 410 students participated in the project. The CO_2 emissions calculation took into account the characteristics of the vehicle fleet of Braga, depending on the percentage of the different fuels used during the year 2017 in Portugal. Whereas in light vehicles there is a clear predominance of gas oil (64.1%), the percentage of gasoline in heavy vehicles is practically insignificant, since diesel is used in 99.49% of these vehicles. In this way, all heavy vehicles were considered to use diesel (100%). On the other hand, DEFRA [17] indicates that the amount of fuel per kilometer (kg/km) consumed by type of vehicle (light and heavy vehicles) and fuel used is 0.07 and 0.06 kg/km for diesel and petrol vehicles, respectively, and 0.24 kg/km for heavy (diesel) vehicles. Based on the number of vehicles circulating in the intervention area, CO_2 emissions from the study area were calculated using Eq. 1.

$$CO_2 \text{emissions} \left(kgCO_2e\right) = \Sigma \left(N_{i,j} \times D_{i,j} \times EF_{i,j}\right) \tag{1}$$

In which:

$N_{i,j}$ – Number of vehicles of category "i" with fuel "j"
$D_{i,j}$ – Distance traveled by vehicles of category "i" with fuel "j" (km)
$EF_{i,j}$ – Emission factor of category "i" vehicles with "j" fuel (kgCO$_2$-eq/km)

The conversion factors for calculating the greenhouse gas emissions considered the parameters set by the British Government [17, 18]. Considering the vehicles fleet composition for light vehicles the emissions factors were 0.17887 kgCO$_2$eq/km and 0.18568 kgCO$_2$eq/km for diesel fuel and gasoline respectively. Regarding heavy vehicles the emissions factors were 0.9339 kgCO$_2$eq/km for heavy-duty diesel vehicles and 1.2194 kgCO$_2$eq/km for buses.

The reduction of CO_2 emissions resulting from "SchoolBus" project was determined considering the daily number of students who used the buses assigned to the project, from the interfaces to the respective educational institutions, in each month since the beginning of the project. In order to facilitate the calculation of the distance traveled by the buses, two centroids were considered: (i) Centroid of the BUILD (CEB) Schools; (ii) Centroid of Schools not BUILD (CENB). The center of the BUILD Schools is the middle center of the bus route that passes through the Dom Diogo de Sousa School, the Leonardo Da Vinci School and the Teresiano School. The centroid of the Schools BUILD is the middle center of the bus route that passes through the School EB2,3 Francisco Sanches, Calouste Gulbenkian School and School EB2,3 André Soares. Knowing the route made by the buses, from the interfaces to the centroids, it was determined the distances traveled by them (Fig. 2).

The reduction of motor vehicle traffic after the implementation of the "SchoolBus" project was obtained by considering that the number of students using the bus as equal to

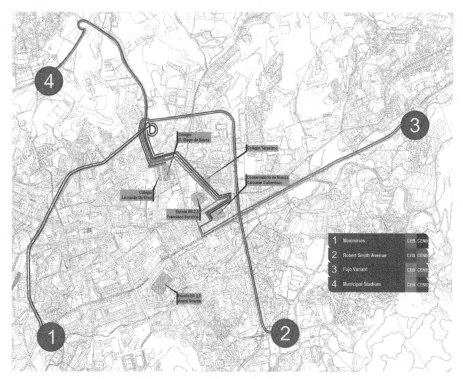

Fig. 2. Routes between interfaces and centroids considered for the calculation of CO_2 emissions (Source: Authors)

the number of private light vehicles that stop circulating in the BUILD area, applying the 1:1 rule. The Table 2 shows the routes traveled by buses and their distances traveled, as well as the number of private vehicles associated with the number of students using the

Table 2. Number of vehicles associated with the number of students using School Bus and the routes and their distances traveled by buses (Source: Authors)

Routes		Travelled distance [Km]	Average number	
Departure	Arrival		Diesel cars	Light cars petrol
Municipal stadium	CENB	10.6	7	4
Municipal stadium	CEB	8.0	15	8
Fojo variant	CENB	6.4	2	1
Fojo variant	CEB	9.4	9	5
Robert Smith avenue	CENB	10.8	2	1
Robert Smith avenue	CEB	8.2	35	19
Maximinos	CENB	11.2	5	3
Maximinos	CEB	8.6	5	3

bus. Finally, through Eq. 1, the contribution of the "SchoolBus" project was calculated for the decarbonization of the BUILD area, which translated into the average annual reduction of 33.8 tonCO$_2$eq.

From the 546 students involved in the awareness actions carried out, 141 students were as well in the sustainable mobility workshops from the Dom Diogo de Sousa School and the Leonardo Da Vinci School (Table 3).

Table 3. Number of stakeholders engaged in each activity (Source: Authors)

Awareness-raising actions		Students engaged
Mobility as a game	Leonardo Da Vinci School	140
	School EB2,3 André Soares	20
	School EB2,3 Francisco Sanches	225
	Calouste Gulbenkian School	20
Sustainable mobility workshops	Dom Diogo de Sousa School - Art students	11
	Dom Diogo de Sousa School - 4th class	108
	Leonardo Da Vinci School	22

One of the awareness-raising actions was a game in which Leonardo Da Vinci school pupils, in groups of 10, had to rank which modes of transport they considered to be more and less sustainable, and which would have the greatest impact on the environment, among several pre-defined modes of transport (walking, cycling, metro, bus, motorcycle and car). This game triggered discussions among the various groups of pupils about the results they had initially considered.

Another carried out action was the game called "Big Mobility" which consisted in answering questions about mobility, in order to address concepts and clarify issues related to the environmental impact created by each mode of transport, involving students and residents in it. This game had already been developed in 2017, by the CMB, during the European Mobility Week. The BUILD project's technical team adapted the idea to the awareness-raising activities carried out, having achieved a participation of 100 students in the "Big Mobility" games.

Students who participated in the Sustainable Mobility workshops indicated that they traveled to school in private cars, as illustrated in Fig. 3. Still, 30% of the total showed a willingness to change to a more sustainable mode of transport, of which 64% mentioned the "SchoolBus" as a choice. On the other hand, 68% of the students indicated that they enjoyed the transportation they currently use to get to school, while 64% traveled by car.

In the awareness actions carried out, it was observed that 50% of the students would like the city to have less road traffic or, in some cases, that it entirely did not exist. The children mentioned the desire to have a city with less air and noise pollution, with fewer road accidents and, in this way, make the city a more attractive place with more green areas (Fig. 4).

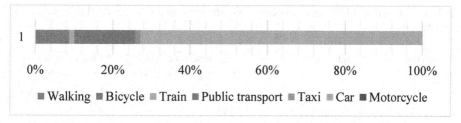

Fig. 3. Modal distribution of students on the commuting to the educational institution [12]

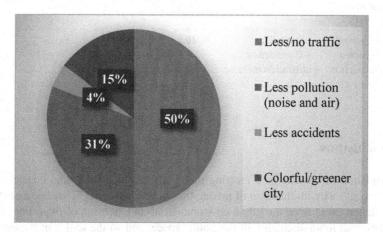

Fig. 4. How the children would like the city to be different from Mental Maps [12]

In the awareness-raising actions there was a potential increase of the CO_2 emissions reduction of the "SchoolBus" project, since 27 new students showed interest in joining the project in the future. Evaluating the impact of these future inscriptions, it was verified that these 27 students would cause a decrease of CO_2 emissions of 7.19 tonCO$_2$eq, which is equivalent to 21% of the current value reached by the "SchoolBus" during this school year (reduction of CO_2 emissions from 33.8 tonCO$_2$eq to 40.9 tonCO$_2$eq).

The "SchoolBus" project, as started in September 2018, may have contributed to its success in achieving large numbers in reducing CO_2 emissions. However, the potential impacts of awareness-raising actions in increasing the reduction of CO_2 emissions allowed to achieve results closer to the maximum values of the range established by the PER Framework method, instead of approaching the minimum values, as shown in Table 4.

The awareness-raising actions carried out made it possible to verify that progressive investments in Sustainable Mobility Measures, through public participation, represent a possibility of greater contribution, even more so when the objective is to achieve the decarbonization goals of a city.

Table 4. Values from PER Framework compared with calculated values of BUILD interventions [12] (Adapted)

Sustainable mobility measure	Intervention	Interventions calculated values [tonCO$_2$eq/year]	PER framework [tonCO$_2$eq/year]	
			Min.	Max.
Corporate, school and personalized mobility plans (Or workplace travel plans)	"SchoolBus" – Current value	**33.8**	33	41
Corporate, school and personalized mobility plans (Or workplace travel plans)	"SchoolBus" – Considering the utilization increase	**40.9**	33	41

4   Conclusions

The correlation of the city as a meeting place, that enabled the social contact of the population get lost with the raise of private transportation modes. That because it kept the population away from the urban environment as an extension of their daily life. It could only lead to an alienation of the public space, and so the scales of the city turn into impersonal relations with the built urban environment.

Considering that, there is a need of increase more active modes of transportation, that are, at the same time, more sustainable modes. This shift can approximate again the population with the city, and consequently it is important to promote a shift from expert technicians into interdisciplinary teams. It will create the possibility of the population to be heard by technical people, at the same time building up a positive relation between society and built environment. That is because it is already clear that the effectiveness of the city planning process is strictly connected with the comprehension by population that the planning is a method able to improve their quality of life.

Furthermore, the Mobility Games as a tool for public participation truly engage the stakeholders and acted as an efficient tool during the phase of motivational increase of the actions. It was a great challenge for the stakeholders to avoid mistakes while answering the questions about sustainable mobility and local traffic laws. At the same time, it was a really ludic activity for young students and helped, in a more functional perspective, in spreading the sustainable mobility concept.

Furthermore, there is much of work in progress, since the impacts of the promoted actions in contribute to a behavioral change regarding sustainable mobility can definitely rise. Once, only 30% of the students could demonstrate the comprehension of co-responsibility on transition to a more sustainable modes of transportation, which means that they also have to change their transportation mode for a more sustainable one, although a great number of the students identified the need of a city with less traffic.

To achieve the goals of the global sustainability agenda not only governmental policies are needed, but indeed the local actions of awareness raising had great impact on behavior, since they were mainly addressed to the young population of the study area. At the same time it was an opportunity to contribute with a transformation of the technicians' perspectives, once approaching the needs of the city-users more closely. Moreover, by creating alternative solutions, such as addressing the Scholar Mobility, that was a great concern on the study area, through the project "SchoolBus", it was possible to actually impact directly issues of urban mobility. Therefore, bringing effective solutions regarding the mobility planning of the city.

Lastly, the confirmation of the continuation of the "SchoolBus" project for the scholar year of 2019/2020 creates an opportunity for future works. It will make possible the further analysis along the months of the raise tendency of students using the service, and thereupon the effective impact calculation of the project. Although the need of continuation of these actions, it is certain that this is already a piece of contribution for the sustainable mobility promotion in the city of Braga.

References

1. United Nations: World Urbanization Prospects: The 2018 Revision (2018)
2. IPCC, Climate Change 2014: Synthesis Report. Contribution of Working Groups I, II and III to the Fifth Assessment Report of the Intergovernmental Panel on Climate Change. Geneva, Switzerland: IPCC (2014)
3. Silva, L.T., Mendes, J.F.G.: City noise-air: an environmental quality index for cities. Sustain. Cities Soc. **4**, 1–11 (2012)
4. Petrovic, N., Bojovic, N., Petrovic, J.: Appraisal of urbanization and traffic on environmental quality (2016)
5. European Union: Sustainable Development in the European Union 2018. Publications Office of the European Union, Luxembourg (2018)
6. European Commission: Next steps for a sustainable European future: European action for sustainability. COM(2016), p. 739 (2016)
7. Korver, W., Stemerding, M., Van Egmond, P., Wefering, F.: CIVITAS Guide for the Urban Transport Professional - Results and Lessons of Long Term Evaluation of the CIVITAS Initiative, Graz, Austria (2012)
8. Wefering, F., Rupprecht, S., Bührmann, S., Böhler-Baedeker, S.: Guidelines. Developing and Implementing a Sustainable Urban Mobility Plan (2014)
9. Gomez Echeverri, L.: Investing for rapid decarbonization in cities. Curr. Opin. Environ. Sustain. **30**, 42–51 (2018)
10. Grandin, J., Haarstad, H., Kjærås, K., Bouzarovski, S.: The politics of rapid urban transformation. Curr. Opin. Environ. Sustain. **31**, 16–22 (2018)
11. Lopez-Ruiz, H.G., Christidis, P., Demirel, H., Kompil, M.: Quantifying the effects of sustainable urban mobility plans, JRC Technical report, vol. EUR 26123 (2013)
12. Souza, P.: People-centered Urban Measures towards Sustainable Mobility, University of Minho (2019)
13. Braga Municipal Council: Plano de Implementação do Braga Urban Innovation Laboratory Demonstrator (BUILD) (2017)
14. ITDP Brasil: Intervenção urbana temporária (Re)pensando a rua em Santana, pp. 1–55 (2018)
15. Poplin, A.: Playful public participation in urban planning: a case study for online serious games. Comput. Environ. Urban Syst. **36**, 195–206 (2012)

16. Pisoni, E., Christidis, P., Thunis, P., Trombetti, M.: Evaluating the impact of 'Sustainable Urban Mobility Plans' on urban background air quality. J. Environ. Manage. **231**, 249–255 (2019)
17. DEFRA: UK Government GHG Convertion Factors for Company Reporting (2017)
18. CTAC: Assessoria técnica no âmbito da avaliação dos impactos decorrentes das ações implementadas, Braga (2018)

Leak Detection in Water Distribution Networks via Pressure Analysis Using a Machine Learning Ensemble

Vivencio C. Fuentes Jr.$^{(\boxtimes)}$ ⓘ and Jhoanna Rhodette I. Pedrasa ⓘ

University of the Philippines, Diliman, 1101 Quezon City, Philippines
vivencio.fuentes@eee.upd.edu.ph, jipedrasa@up.edu.ph

Abstract. Water distribution networks (WDNs) are vital infrastructure which serve as a means for public utilities to deliver potable water to consumers. Naturally, pipelines degrade over time, causing leakages and pipe bursts. Damaged pipelines allow water to leak through, incurring significant economic losses. Mitigating these losses are important, especially in areas with water scarcity, to allow consumers to have adequate water supply. Globally, as the population increases, there is a need to make water distribution efficient, due to the rising demand. Thus, leak detection is vital for reducing the system loss of the network and improving efficiency.

Monitoring WDNs effectively for leakage is often a challenging task due to the size of the area it covers, and due to the need to detect leaks as early as possible. Traditionally, this is done via pipeline inspection or physical modeling. However, such techniques are either time-consuming, resource intensive, or both. An alternative is machine learning (ML), which maps the relationship between pipeline data to detect leakages. This allows for a faster, yet reasonably accurate model for detection and localization. Machine learning techniques could be utilized together as an ensemble, which allows these techniques to work in conjunction with each other. Wavelet decomposition will be performed on the data to allow for a smaller dataset, as well as utilizing possible hidden features for the machine learning model.

Keywords: Water distribution networks · Leak detection · Machine learning

1 Introduction

1.1 Water Distribution Networks

Water distribution networks (WDNs) are systems designed and implemented to deliver potable water from a source to a consumer. However, pipeline

This work is supported by a Commission on Higher Education (CHED) Philippine California Advanced Research Institues (PCARI) grant under Project IIID54 - Resilient Cyber Physical Societal Scale Systems.

P. Pereira et al. (Eds.): SC4Life 2019, LNICST 318, pp. 31–44, 2020.
https://doi.org/10.1007/978-3-030-45293-3_3

infrastructure gradually experience deterioration, which could be due to natural aging, environmental damage, or unauthorized human interference [1]. Damaged pipelines allow water to leak through, which would cause significant economic losses. As the world population grows, there is a need for distributing water efficiently due to scarcity in water supply. Mitigating losses is key, especially in areas with water scarcity, making leak detection an important part for improving the efficiency of a network.

1.2   Leak Detection

Without using automated sensing or analytical methods, leakages are traditionally reported via water utility personnel examining above-ground meters along the WDN [2]. Due to the nature of human involvement, this is a time-consuming method that could possibly incur significant losses before it is addressed.

As leakage events follow hydraulic principles, sensors could be used to monitor characteristic changes within the pipeline. Pressure drops and flow imbalance in certain areas could imply disruption in normal pipeline operations [1].

Data acquired from sensors in pipelines could be studied to develop analytical models. By combining known hydraulic equations and state analysis of the pipeline, physical modeling could be performed for a network [3]. Further analyses of data could be done by applying machine learning techniques to establish correlation between different factors, and distinguish false positives from actual leaks [4]. More often than not, a combination of certain aspects of these techniques are used to improve accuracy of leak detection [5].

1.3   Data Analysis

Datasets containing real-world data could be used to model existing water distribution networks and possible leakage occurrences in conjunction with actual demand patterns [6]. However, in the absence of real-world data, EPANET software would be used to simulate leakage in water distribution networks [7]. Components such as pipes and valves could be implemented in simulations to reflect physical models. Datasets are constructed from such simulations, and could be used for analysis. For this study, the datasets are generated with emphasis on varying sensor density and topology. Datasets from these networks would then be evaluated to test the machine learning model. Using machine learning, leakage detection in these networks could be performed [8].

1.4   Overview

The main objective of this research is to perform simulations on different water distribution networks, and develop a machine learning approach for leak detection. Factors affecting WDN operations such as water demand and physical characteristics of the WDN will be taken into consideration. Performance of the machine learning model will be evaluated on different network topologies and sensor density.

2   Related Work

2.1   Leak Detection Methods

Without the usage of other leak detection techniques, on-site inspections by water utility personnel are relied upon [2]. Above-ground sensor reading equipment must be deployed along the WDN for a technician or engineer to be able to note sensor data. As the response highly involves human intervention, the system could possibly undergo significant losses before the appropriate action is implemented [1].

Acoustic sensors could be installed along the WDN to detect sound signals, which could be used with signal processing techniques to detect and localize leakages [2]. Signals would be higher in amplitude near the leaks, allowing for localization. This method has the advantage of not needing a complex mathematical model. However, acoustic sensing techniques are easily affected by the physical environment around it; external noise might interfere with sensor readings [9]. The cost of implementing the hardware for the system would be high as well, especially for larger networks [1].

Another technique used for leak detection are balancing methods. Under normal operation, input and output mass flow rate of each pipeline are equal [2], which is supported by the law of conservation of mass. Thus, imbalances of these metrics would imply leakage within the pipeline. The balance between input and output flows could be represented by the equation

$$M_I - M_O = \Delta M_{pipe} \tag{1}$$

where M_I and M_O represent mass in the inlet and outlet of the pipe, respectively, while ΔM_{pipe} represents the imbalance between the two. As mass is proportional to the flow rates, input and output flows could also be represented by the equation

$$Q_I - Q_O = \Delta Q_{pipe} \tag{2}$$

where Q_I and Q_O similarly represent flow rates in the inlet and outlet of the pipe, respectively, while ΔQ_{pipe} represents the imbalance between the two [1]. To be within range of possible errors in sensor readings, an alarm threshold is implemented based on pressure or flow values. While rather simplistic in theory, utilizing pressure and flow data is a fundamental basis for techniques that rely on data processing of hydraulic models.

Transient analysis could also be utilized as a leak detection method [10]. Transient information generated by leakage scenarios could be extracted and then evaluated accordingly. The main idea of transient analysis is to compare signals received by a sensor in a leakage event to signals under normal operation. Focusing on the transient signals can offer more information rather than simply monitoring the WDN at its steady state [10,11].

Model-based methods could be assessed, to determine the state of subsections of a WDN based on water hammer equations [3]. The underlying fundamental principle relating flow and pressure could be summarized in equations

$$\frac{\partial H}{\partial t} = -\frac{c^2}{gA}\frac{\partial Q}{\partial z} \tag{3}$$

$$\frac{\partial Q}{\partial t} = -Ag\frac{\partial H}{\partial z} - \frac{fQ|Q|}{2DA} \tag{4}$$

where t and z represent coordinates in time and space, respectively, H representing pressure, c representing wave speed, g representing Earth's gravity, Q representing flow rate, D representing the diameter of the pipeline, A representing the cross-sectional area of the pipe, and f representing the friction coefficient [3].

While model-based methods based on physical properties of a WDN yield high accuracies in fault detection, it is not feasible to use in large-scale networks due to the amount of parameters and data involved [12,13]. An alternative is to use machine learning techniques, which generate similarly high accuracies, but require less computational load [8,14]. Machine learning maps the dependent and independent parameters of a given system, with little prior process knowledge. Performance of machine learning techniques are based on the design of the technique, which could be improved over time with data [13].

Machine learning techniques and their properties could be incorporated into a single learning algorithm called an ensemble method [15]. Ensemble methods are multiple machine learning techniques that are trained cooperatively to produce better predictive performance. In the field of hydrology, there have been proposals to utilize ensemble methods to improve the performances of WDNs [14,16]. Pairing up single methods with each other could help decrease unwanted levels in bias and variance, which could ultimately increase predictive performance.

2.2   EPANET

EPANET is a simulation software for modeling water distribution networks and examine hydraulic behavior in pipe networks. It is able to simulate different WDN entities, such as pipes, tanks, valves, and reservoirs. Factors such as water pressure, water flow, and chemical concentrations could be monitored within the system. EPANET also provides a visual simulator to help the user build pipe networks and edit properties such as pipe diameter or valve function [7].

By simulating emissions within a water distribution network, EPANET is able to model leakage events, and the properties of the network during such an event could be studied. EPANET has been used in modeling and detecting leakages in WDNs of real-life communities and small-scale testbeds [17,18] (Fig. 1).

EPANET takes advantage of combining demand patterns and leakage properties to model leaks within a specified time frame. This would help in detecting the possibility of leakage within the WDN [18]. A relation used in EPANET to model leakage is as follows [17]:

$$Q = E_c * P^{P_{exp}}, \tag{5}$$

where Q represents flow, E_c represents the emitter coefficient, P represents pressure, and P_{exp} represents the pressure exponent, which is usually set to 0.5 for water networks.

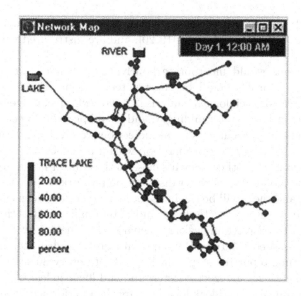

Fig. 1. EPANET window simulation

3   Methodology

The methodology is composed of three major tasks: simulations, feature extraction, and classification. Simulation modeling includes the construction of a functional and realistic water distribution network in simulation software. Simulations involving leakage events are carried out afterwards, done on WDNs with varying topology and sensor density.

The proposed solution is to apply wavelet series decomposition on pressure data, and to use the energy of each wavelet as a feature for the machine learning model. Specifically, a CNN ensemble is implemented as part of the machine learning model to analyze the data. The CNN ensemble would effectively classify the pressure profiles from the sensors as to whether or not they have a leak. Instead of using each recording from each sensor node as a feature, the energy of each wavelet is used, effectively reducing the size of the dataset. This allows the machine learning model to learn from another feature, which could result in better detection.

3.1   Hydraulic Simulations

In the absence of datasets for real-life water distribution networks, simulations are performed in EPANET to produce data from networks with similar characteristics. Water distribution networks are constructed in EPANET by implementing elements such as pipes, junctions, and reservoirs. These networks in the EPANET program reflect certain parts, if not the whole, of actual WDNs.

For the analysis, three distinct EPANET networks are used: the Cherry Hills/Brushy Plains network, the New York City Tunnel network, and the

Fossolo network. These networks are derived from actual implementations of existing water distribution networks [19,20], and are distributed with EPANET releases [7].

The simulations would incorporate hourly water demand profiles used by households [18]. A profile could be constructed via analysis of previous consumption data [21], or sociological analysis [22]. For example, usage in residential areas is typically at lowest after midnight, and would peak during early mornings or early evenings [23]. Accuracy of leak detection would depend on how much volume of water is taken into consideration within pipelines.

Pipes, junctions, and other elements close to the emitter nodes are monitored for changes. The model for the water distribution networks is able to capture key characteristics, and these will be the basis for leak detection. Such characteristics include pressure, water flow, water head, and water quality. Out of these, pressure changes are most reflective of leakage scenarios [4], making it the ideal data input for the proposed solution. Pressure data are recorded accordingly based on EPANET's time reporting step, which was set to every 30 s. Changes in the pressure profile is then analyzed to determine and localize leaks.

Changes on leak sizes and leak locations are then performed. These combinations of changes are used to create datasets for machine learning. To introduce leakage events, the emitter coefficients of the nodes within the network are be increased. Shown in Fig. 2 are the differences between the pressure profile of a node in the Cherry Hills/Brushy Plains network within a 24-h period upon increasing the emitter coefficient. The drop in pressure is found to be consistent with hydraulic principles.

To facilitate leak detection, sensor nodes are implemented into every EPANET simulation of a water distribution network. These are assigned out of existing junction nodes, and the placement of these would have an impact in detecting leaks. This allows for a mechanism to monitor and record pressure profiles within the network, which would then be analyzed accordingly [18].

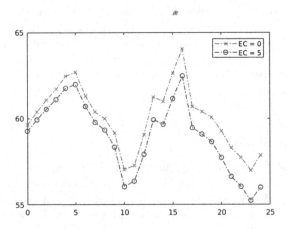

Fig. 2. Pressure profile of the same node at $E_c = 0$ and at $E_c = 5$

To test the effect of sensor density, sensor placement is done differently on each simulation. Specifically, different simulations have sensors that are placed on locations that are 1-, 2-, and 3-hop distances from non-sensor junction nodes, without overlap.

3.2 Feature Extraction and Wavelet Decomposition

The data streams that will be observed are pressure streams coming from sensor nodes. Intuitively, pressure data observed from components away from the leakage event would be inversely proportional to the distance from the leak. Generally, leakage within the vicinity would result in more drastic pressure drops.

Furthermore, wavelet transforms could be applied to these signals, where they would be decomposed into their respective approximation coefficients and detail coefficients [4,24]. The approximation coefficient corresponds to the lower frequency of the original signal, while the detail coefficients correspond to the higher frequency of the original signal. Different levels of the detail coefficient could possibly detect transients in the signal, which could be correlated with detecting leakages in the pipeline, as leakage events come with changes of pressure. The detail coefficients also capture the noise element of the signal, while the approximation coefficient resemble the original signal (Fig. 3).

The energy of the coefficients would be computed, and used as the feature vectors for each sensor, effectively reduceing the input size of the dataset. Utilizing the energy of the pressure signals instead of raw pressure data allow for minimizing energy consumption of the sensors in sending data over a network [18], as well as allowing the CNN ensemble to possibly find more hidden features [4].

These data would then be analyzed via machine learning by a CNN ensemble. It is expected that the different networks would provide different pressure profiles for the CNN ensemble to learn.

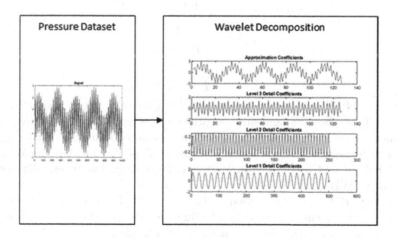

Fig. 3. Wavelet Decomposition

3.3   Classification via Machine Learning Ensemble

The main concern of the study is to identify the occurrence of leakage scenarios given a profile of a water distribution network. The study also aims to compare the effectiveness of the proposed solution on different network topologies and different sensor densities.

A machine learning ensemble is implemented by training multiple machine learning sub-models then stacking them together to form an ensemble. Specifically, an ensemble consisting of a stacked convolutional neural network (CNN) sub-models is used to classify leakage events by learning patterns from the generated wavelets. A CNN is a machine learning implementation used for feature extraction [25] and has been used extensively in the field of hydrology [8,26]. This could be utilized to reveal hidden features relevant for leak detection. The CNN ensemble would act as a binary classifier for leak detection, where its output would be to determine whether or not a leak exists within the network. Adjustments in configurations are made to maximize performance for a specific network model.

Data generated from EPANET simulations are used as input for the machine learning model. The models are implemented in Python using Keras and Tensorflow as framework, with a 70/30 split of the dataset for the training and test set.

3.4   Evaluation Metrics

The techniques above will be evaluated with respect to certain metrics: classification accuracy, true positive rate (TPR), false positive rate (FPR), and the area under the curve (AUC) of the receiver operating characteristic (ROC) [17,27]. A summary of these metrics is as follows:

$$Accuracy = \frac{correct}{total} * 100 \tag{6}$$

$$TPR = \frac{TP}{TP + FN} \tag{7}$$

$$FPR = \frac{FP}{FP + TN} \tag{8}$$

where TP, TN, FP, and FN represent the total amount of true positives, true negatives, false positives, and false negatives respectively. A true positive (TP) is defined as correctly identifying a leakage scenario occurring in the WDN, while a true negative (TN) is defined as correctly identifying that the WDN is operational without leaks. A false positive (FP) is defined as incorrectly noting a leak in the WDN.

The AUC-ROC is an often used evaluation metric for machine learning [27], which is generally characterized by plotting the true positive rate against the false positive rate. The area is a measure of how much the model is able to distinguish between classes. A higher AUC corresponds to a better classifier.

4   Results and Discussion

Shown in Table 1 are the physical attributes of the tested networks. These serve as a baseline on how network topologies physically differ from one another.

Table 1. Physical characteristics of different network topologies

Metric	New York tunnel	Cherry plains	Fossolo
Nodes	19	34	36
Avg. Node Connections	2.105263	2.294118	3.194444

Tables 2, 3, 4 present the performance of the proposed leak detection method. The ensemble CNN method was able to reach accuracies of more than 90% in each topology, even when considering the differences in sensor density. While this is the case, the accuracy of the model is dependent on how complex or large the network; leak detection on networks with more connections between nodes tend to be slightly less accurate than networks with fewer connections. Similarly, the true positive rate for all topologies and sensor densities reach an adequate percentage for leak detection. This metric is relevant as there is a need to correctly determine the presence of leaks in the event of one occurring. In all region sizes, it is seen that the Fossolo network has the lowest true negative rate, while the Cherry Plains network has the highest. The true negative rate of the Fossolo network is at its lowest when using data from the 3-hop node distance region. This may be due to the fact that the dataset is mostly comprised of leakage scenarios, with non-leakage scenarios only taking up roughly 14% of the datasets, and compounded further by having generated less data from a network configuration with less sensors. It is likely that the CNN ensemble fails to find more hidden features in these networks, where the pressure profiles are comprised of more nodal connections; hence the low true negative rate. Generally, this is more acceptable than the opposite, where there is a low true positive rate, and a high negative rate, as inaction for an existing leak would entail more economic losses than routinely checking on pipe operations.

Table 2. Performance of the model for leak detection of different network topologies, with the sensor density based on 1-Hop node distances

Sensor density based on 1-Hop node distances			
Metric	New York tunnel	Cherry plains	Fossolo
Regions	8	14	10
Accuracy	0.987737584	0.98560658	0.96956944
TPR	0.994318182	0.98458498	0.990661188
TNR	0.946188340	0.992268041	0.832524272

Table 3. Performance of the model for leak detection of different network topologies, with the sensor density based on 2-Hop node distances

Sensor density based on 2-Hop node distances			
Metric	New York tunnel	Cherry plains	Fossolo
Regions	4	6	4
Accuracy	0.984671980	0.977039068	0.966655876
TPR	0.994318182	0.973913043	0.988793425
TNR	0.923766816	0.99742268	0.822815534

While there is not much difference in the accuracy and true positive rate metrics between sensor densities, a noticeable trend is seen in the true negative rate. As the sub-regions of the network become larger, a decline in the true negative rate can be noticed. This is corroborated by the decrease in the AUC-ROC for the Fossolo network, as shown from Figs. 4, 5, 6, as it is affected by the increase in false negatives. It is possible that covering a large area within the

Table 4. Performance of the model for leak detection of different network topologies, with the sensor density based on 3-Hop node distances

Sensor density based on 3-Hop node distances			
Metric	New York tunnel	Cherry plains	Fossolo
Regions	3	4	3
Accuracy	0.984671980	0.979095271	0.946260926
TPR	0.995028409	0.980237154	0.987672768
TNR	0.919282511	0.971649485	0.677184466

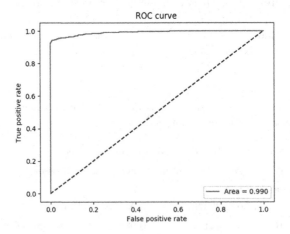

Fig. 4. ROC curve of the Fossolo network, 1-hop region size

WDN makes it more difficult for the machine learning model to learn connections in the pressure profile and correctly classify non-leakage scenarios. Having more connections on average for each junction further exacerbates this problem, as nodal pressures are more affected by neighboring nodes.

In general, the model was able to correctly classify leakage and non-leakage scenarios correctly. From the results gathered, significant increases in accuracy, TPR, and TNR for leak detection can be achieved by allowing the model to be trained further, as each network configuration was trained using the same machine learning model and the same amount of training periods. Networks with higher connections between nodes require more training as well, to find more connections between the different pressure profiles generated by each node.

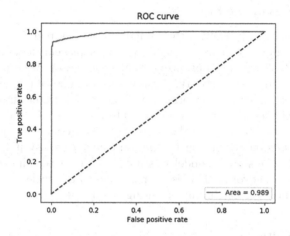

Fig. 5. ROC curve of the Fossolo network, 2-hop region size

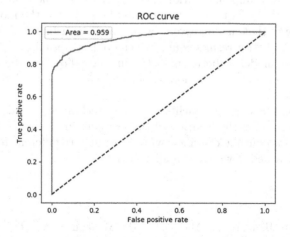

Fig. 6. ROC curve of the Fossolo network, 3-hop region size

5   Conclusions

This study explored the problem of detecting leakages in water distribution systems. The proposed solution of using a CNN ensemble with wavelet decomposition was able to determine occurrences of leaks within the pipeline. Wavelet decomposition allowed the data used for machine learning features to be relatively small, while retaining information regarding changes in the pressure profile. By using a CNN as a classifier, connections in the pressure profiles were analyzed and yielded sufficient performance by the model. Generally, the model is able to perform well on different network topologies, different sensor densities, and certain combinations thereof.

5.1   Limitations of the Study

The research dealt exclusively with simulated data, as real-life datasets were found to be incomplete or hard to come by. Sensor placement within the simulated networks were also not optimized. While the sensors were physically distributed equitably over the network, a study was not done on how specific sensor placements affect the overall performance of the model. Determining where the leak is located is likewise not incorporated into the study. As the study simply aims to detect whether or not a leak exists in the system, the utility of this knowledge depends on the size of the network and how easy it is for the leak to be alleviated. The trained model was not used in actual implementations of water distribution networks. This is primary due to the scale of the simulated networks, and the difficulty in implementing small-scale ones.

5.2   Future Work

Improvements in the implementation of the CNN ensemble could still lead to better performance. From the results, it could be seen that the model performs worse on networks that are relatively large and have a higher number of connections per node; better results could be achieved by optimizing the layers and parameters of the model. A better model would also allow for a better classifier, allowing for the classification of more regions within the network or more dense sensor distributions.

Similar to leak detection, machine learning techniques could also possibly be used for determining the location of the leak within the network. The CNN could be used as a multiple classifier, where outputs are different regions within the network, narrowed down to a smaller area.

References

1. Adedeji, K.B., Hamam, Y., Abe, B.T., Abu-Mahfouz, A.M.: Towards achieving a reliable leakage detection and localization algorithm for application in water piping networks: an overview. IEEE Access **5**, 20272–20285 (2017). https://doi.org/10.1109/access.2017.2752802

2. Wan, J., Yu, Y., Wu, Y., Feng, R., Yu, N.: Hierarchical leak detection and localization method in natural gas pipeline monitoring sensor networks. Sensors **12**(12), 189–214 (2011). https://doi.org/10.3390/s120100189

3. He, Y., Li, S., Zheng, Y.: Distributed state estimation for leak detection in water supply networks. IEEE/CAA J. Autom. Sin. **PP**(99), 1–9 (2017). https://doi.org/10.1109/JAS.2017.7510367

4. Gupta, G.: Monitoring Water Distribution Network using Machine Learning. KTH Royal Institute of Sweden, Stockholm (2017)

5. Xu, Q., Liu, R., Chen, Q., Li, R.: Review on water leakage control in distribution networks and the associated environmental benefits. J. Environ. Sci. (China) **26**(5), 955–961 (2014). https://doi.org/10.1016/S1001-0742(13)60569-0

6. Conejos, M.P., Alzamora, F.M., Alonso, J.C.: A water distribution system model to simulate critical scenarios by considering both leakage and pressure dependent demands. Procedia Eng. **186**, 380–387 (2017). https://doi.org/10.1016/j.proeng.2017.03.234

7. EPANET. https://www.epa.gov/water-research/epanet. Accessed 23 May 2018

8. Kang, D., Lansey, K.: Novel approach to detecting pipe bursts in water distribution networks. J. Water Resour. Plan. Manage. **140**(1), 121–127 (2014). https://doi.org/10.1061/(ASCE)WR.1943-5452.0000264

9. Fang, C.M., Lin, S.C.: 5.4 GHz high-Q bandpass filter for wireless sensor network system. In: IEEE SENSORS 2009 Conference (2009). https://doi.org/10.1109/ICSENS.2009.5398458

10. Colombo, A.F., Lee, P., Karney, B.W.: A selective literature review of transient-based leak detection methods. J. Hydro-Environ. Res. **2**(4), 212–227 (2009). https://doi.org/10.1016/j.jher.2009.02.003

11. Wang, X., Ghidaoui, M.S.: Identification of multiple leaks in pipeline: linearized model, maximum likelihood, and super-resolution localization. Mech. Syst. Signal Process. **107**, 529–548 (2018). https://doi.org/10.1016/j.ymssp.2018.01.042

12. Quiñones-Grueiro, M., Verde, C., Llanes-Santiago, O.: Demand model in water distribution networks for fault detection. IFAC-PapersOnLine **50**(1), 3263–3268 (2017). https://doi.org/10.1016/j.ifacol.2017.08.460

13. Goyal, M.K., Ojha, C.S.P., Burn, D.H.: Machine learning algorithms and their application in water resources management. In: Sustainable Water Resources Management, pp. 165–177. American Society of Civil Engineers (2017). https://doi.org/10.1061/9780784414767.ch06

14. Kang, J., Park, Y., Lee, J., Wang, S., Eom, D.: Novel leakage detection by ensemble CNN-SVM and graph-based localization in water distribution systems. IEEE Trans. Ind. Electron. **65**(5), 4279–4289 (2018). https://doi.org/10.1109/TIE.2017.2764861

15. Dietterich, T.G.: Ensemble methods in machine learning. In: Kittler, J., Roli, F. (eds.) MCS 2000. LNCS, vol. 1857, pp. 1–15. Springer, Heidelberg (2000). https://doi.org/10.1007/3-540-45014-9_1

16. Shi, F., Liu, Z., Li, E.: Prediction of pipe performance with ensemble machine learning based approaches. In: 2017 International Conference on Sensing, Diagnostics, Prognostics, and Control (SDPC), Shanghai, pp. 408–414 (2017). https://doi.org/10.1109/SDPC.2017.84

17. Nasir, M.T., Mysorewala, M., Cheded, L., Siddiqui, B., Sabih, M.: Measurement error sensitivity analysis for detecting and locating leak in pipeline using ANN and SVM. In: 2014 IEEE 11th International Multi-Conference System Signals Devices, SSD 2014, pp. 7–10 (2014). https://doi.org/10.1109/SSD.2014.6808847

18. Gupta, K., Kishore, K., Jain, S.C.: Modeling and simulation of CEERI's water distribution network to detect leakage using HLR approach. In: 2017 6th International Conference on Reliability, Infocom Technologies and Optimization (ICRITO) (Trends and Future Directions) (2017). https://doi.org/10.1109/ICRITO.2017.8342440

19. Rossman, L.A., Clark, R.M., Grayman, W.M.: Modeling chlorine residuals in drinking-water distribution systems. J. Environ. Eng. **120**(4), 803–820 (1994). https://doi.org/10.1061/(ASCE)0733-9372(1994)120:4(803)

20. University of Exeter Centre for Water Systems. http://emps.exeter.ac.uk/engineering/research/cws/resources/benchmarks/. Accessed 23 May 2018

21. Delorme-Costil, A., Bezian, J.J.: Forecasting domestic hot water demand in residential house using artificial neural networks. In: 2017 16th IEEE International Conference on Machine Learning and Applications (ICMLA), Cancun, pp. 467–472 (2017). https://doi.org/10.1109/ICMLA.2017.0-117

22. Domene, E., Sauri, D.: Urbanisation and water consumption: influencing factors in the metropolitan region of Barcelona. Urban Stud. **43**(9), 1605–1623 (2006). https://doi.org/10.1080/00420980600749969

23. Letting, L.K., Hamam, Y., Abu-Mahfouz, A.M.: Estimation of water demand in water distribution systems using particle swarm optimization. Water, 9(8) (2017). https://doi.org/10.3390/w9080593

24. Fagiani, M., Squartini, S., Severini, M., Piazza, F.: A novelty detection approach to identify the occurrence of leakage in smart gas and water grids. In: Proceedings of the International Joint Conference Neural Networks, vol. 2015, September 2015. https://doi.org/10.1109/IJCNN.2015.7280473

25. Wu, Z.Y., El-Maghraby, M., Pathak, S.: Applications of deep learning for smart water networks. Procedia Eng. **119**, 479–485 (2015). https://doi.org/10.1016/j.proeng.2015.08.870

26. Xu, Y., Zhang, J., Long, Z., Chen, Y.: A novel dual-scale deep belief network method for daily urban water demand forecasting. Energies **11**(5), 1068 (2018). https://doi.org/10.3390/en11051068

27. Sokolova, M., Lapalme, G.: A systematic analysis of performance measures for classification tasks. Inf. Process. Manage. **45**(4), 427–437 (2009). https://doi.org/10.1016/j.ipm.2009.03.002

Solutions for Improving the Energy Efficiency of Buildings Refurbishment

Maria de Fátima Castro, Joana Andrade, Catarina Araújo, and Luís Bragança[✉]

CTAC, University of Minho, Guimarães, Portugal
info@mfcastro.com

Abstract. The buildings sector contributes to 30% of annual greenhouse gas emissions and uses about 40% of energy. In this scenario and regarding the sustainable concerns, different strategies and directives have been developed. However, this consumption can be reduced by between 30% and 80% through commercially available technologies.

A critical and evolutionary way of thinking about the energy (and other resources) demand, management, and supply is necessary, because there is a clear concern about irreversible impacts to the world and a scarcity of the resources as well. Energy supplies should be mostly or entirely through renewable resources and highly efficient technologies put in place to achieve solution such as nearly Zero Energy Buildings (nZEB). The strategies and the rehabilitation benefits are more and more recognised, used and increasingly considered by the stockholders. So, the aim of this article is to: critical analyse the state-of-art regarding this matter; discuss the existing solutions, directives, and strategies; and present a case study where the energy performance and the economic viability are discussed. The case study is a Portuguese building model which represents the conventional construction between 1960 and 1990. On it, three scenarios of refurbishment were tested, and their benefits and costs are presented.

Keywords: Energy efficiency · nZEB · Building rehabilitation · Building retrofit

1 Introduction

The connection between climate change and carbon emissions is impossible to decouple, since emissions are strongly linked to energy production and use. Measures are therefore being taken to reduce energy use, promote energy efficiency and increase the use of renewable energy. Improving the energy performance of buildings is one of the most economically viable measures to achieve European climate change objectives and to stimulate sustainable growth [1]. This not only leads to environmental benefits, but also promotes important social and economic benefits by reducing energy poverty, improving thermal comfort and indoor air quality, improving health and productivity, creating jobs, and promoting finance [2].

Buildings are responsible for about 40% of the total energy use in the European Union (EU) and for 36% of total CO_2 emissions. Furthermore, it is inside buildings

© ICST Institute for Computer Sciences, Social Informatics and Telecommunications Engineering 2020
Published by Springer Nature Switzerland AG 2020. All Rights Reserved
P. Pereira et al. (Eds.): SC4Life 2019, LNICST 318, pp. 45–56, 2020.
https://doi.org/10.1007/978-3-030-45293-3_4

that people spend 90% of their life [3]. It is estimated that in the EU, Switzerland and Norway, there are about 25 billion square meters of constructed floor space. So, if the entire constructed area were to be concentrated, the occupied land would be equivalent to the territorial area of Belgium – 30,528 km^2 [4]. On the other hand, it is known that 75% to 90% of the existing building will still be in use by 2050. Much of it was built before 1990 (before EU legislation) and is considered energetically inefficient. So, it is necessary to increase the energy rehabilitation rate by at least 3% to achieve the objectives of the Paris Agreement and the European Energy Efficiency Directive [5].

1.1 Methodology and Objectives

This paper intends to present a critical analysis of the state of the art, discussing existing solutions, directives, and strategies with the aim of improving the energy efficiency of the existing building and presenting a case study about its economic viability. For this, a review of the most relevant literature on the subject, namely the EU directives and concepts defined by them, was carried out. Finally, five rehabilitation scenarios and their economic viability were tested in a case study of a conventional building model of a Portuguese building between the 1960s and 1990s.

2 Strategies and Directives for Improving Energy Efficiency

Considering the need to reduce energy consumption and improve its efficiency, the European Union released the Directive 2002/91/EC, known as EPBD - Energy Performance of Buildings Directive, which has since been reformulate by Directive 2010/31/EU, recognised as EPBD-recast, and amended by Directive (EU) 2018/844.

2.1 Directive of Energy Performance of Buildings

The Directive 2002/91/EC, approved in December of 2002, aimed to promote the improvement of the energy performance of buildings in the EU by means of economic feasibility, considering the climate and the local conditions of each member state (MS). The purpose was to increase the energy efficiency of building and thereby improve their quality (new buildings or renovated), reduce the external energy dependency, decrease the emission of greenhouse gas (GHG) and increase the population awareness and information. The EPBD imposed the energy certification system to demonstrate their performance level, through an integrated calculation method. This method should account for thermal characteristics of the building, heating and cooling systems, domestic hot water (DHW) preparation, ventilation, lighting, or passive solar systems, among others. The EPBD imposed also minimum requirements for the energy performance of new buildings and subject to major renovation works, and the need to regularly inspect boilers and air-conditioning systems and the heating system at each 15 years [6].

This directive implementation had some obstacles such as the MS diversity of the built environment and the low ambitions in some. The low renovation rate was also responsible for aggravating compliance with the objectives, as well as the lack of credibility of some energy certificates and the of the obligation to report the national implementation results.

In 2010, the EPBD was reformulated by the Directive 2010/31/EU (EPBD-recast). This new directive intended to: (i) reduce the CO_2 emissions to mitigate the climate change and, (ii) promote the development of sustainable and energy efficient solutions. Therefore, the following goals were established until 2020: (i) 20% reduction in energy, (ii) 20% reduction in CO_2 emissions and, (iii) 20% increase in use of renewable energy.

This requires the MS to establish minimum energy requirements considering cost-optimal levels and to revise their energy standards in building regulations, at regular intervals, which shall not be longer than five years. The EPBD-recast also established the concept of nZEB – nearly zero energy buildings – for new buildings from 2019 (public sector) and 2021 (all new buildings). The cost-optimal levels of the minimum energy performance requirements mean the energy performance level which leads to the lowest cost during the estimated economic lifecycle, see Fig. 1 [7].

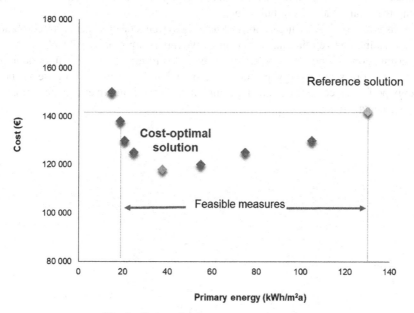

Fig. 1. Cost-optimal concept representation.

As a result of the reduction of the minimum energy requirements, the maximum allowed values, as for instance, the thermal transmission coefficient (U – W/m^2 °C), were lowered regarding the requirements imposed by Directive 2002/91/EC. In Portugal, the U-value for external walls decreased from 1.4 W/m^2 °C to 0.4 W/m^2 °C and from 0.9 W/m^2 °C to 0.35 W/m^2 °C for roofs.

All MS were compelled to implement national plans for energy efficiency or policies with alternative measures that enable reducing the final energy. The decrease of energy consumption after the implementation of EPBD and EPBD-recast can be seen in Fig. 2.

The mean annual final energy consumed which decreased 2.1 kWh/(m²·year) until 2006, is now decreasing at higher rate of 3.8 kWh/(m²·year).

Recently, Directive 2018/844 of 30 May 2018 amending Directive 2010/31/EU on the energy performance of buildings and Directive 2012/27/EU on energy efficiency was published. The main goal of this new directive is to increase the average renovation rate in a cost-effective manner, i.e., (i) introduce building automation and control as an alternative to physical inspections; (ii) encourage the implementation of efficient mobility infrastructures and, (iii) introduce a smart readiness indicator to measure the technologic capacity of buildings. Thus, the following amends can be highlighted [5]:

- Insertion of new definitions as "building automation and control system";
- Implementation, until 2050, of a long-term renovation strategy to support the renovation of the building stock, into a highly energy efficient and decarbonised one;
- Convene the Commission to the implement of an optional common Union scheme for rating the smart readiness of buildings;
- Establishment of regular inspections of heating systems or of systems for combined space heating and ventilation, with an effective rated output of over 7 kW;
- Determination of the primary energy use in kWh/(m²·year) as a numeric indicator to express the energy performance of a building for the purpose of both energy performance certification and compliance with minimum energy performance requirements.

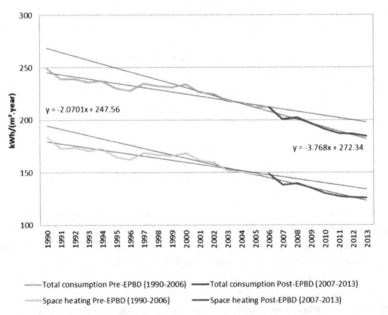

Fig. 2. Energy use evolution in the residential sector, the EU between 1990 and 2013 [8].

2.2 nZEB – nearly Zero Energy Buildings

The nZEB concept arose with the EPBD-recast. This concept refers to buildings that have a very high energy performance (Directive 2010/31/EU, 2010). According to the Directive definition the amount of energy required should be "nearly zero or very low" and "should be covered to a very significant extent by energy from renewable sources, including energy from renewable sources produced on-site or nearby". Nevertheless, none precise definition was established, and each MS should draw up their definition, considering the following:

- Period – time rate used for the energy balance (annual, monthly, and weekly);
- Boundary – on-site and nearby system boundary (building, site, neighbourhood, …);
- Weighting system – which weighting factors to consider in the energy balance, dynamic or static;
- Energy flow – which use to consider (heating, cooling, lighting, DHW supply, …).

Therefore, it is of major importance to establish a holistic nZEB approach, considering not just the reduction of energy use in the building through solar passive design measures, but also satisfying the remaining needs with energy from renewable sources (Fig. 3).

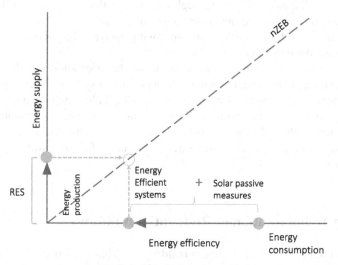

Fig. 3. nZEB holist approach.

3 Existing Buildings and Rehabilitation

It is essential to stimulate and promote the rehabilitation of the existing residential buildings, considering the European objectives already presented. Non-residential buildings, on the other hand, account for about 25% of the European built constructions, which is a more heterogeneous sector and therefore more complex than the residential sector. Additionally, according to an analysis of energy efficiency certificates in Europe, about 97.5% of the building constructions should be improved so that it could be made up of highly efficient buildings, and thus the objective of decarbonisation until 2050 will be reached [4].

3.1 Directive of Energy Performance of Buildings

The concept of rehabilitation is defined as an intervention, more or less extensive, necessary to do in a building or property, aiming to increase its useful and service life, its economic value, the quality of life of the inhabitants and the implementation of good practices of energy efficiency [9]. In this way, the thermal and energy rehabilitation of buildings is imperative as it allows correcting of other existing anomalies in buildings, as well as facilitating the compliance with the existing directives and the reduction of the energy needs, thus reducing the use of energy and emissions of GHG [10]. Therefore, rehabilitation allows implementing measures of energy efficiency and sustainability, providing the increase of thermal and acoustic comfort, and indoor air quality.

The thermal and energy rehabilitation of a building can act on several fronts, such as:

- The opaque building envelope - through the introduction of thermal insulation for correction of thermal bridges and reduction of heat losses;
- The non-opaque building envelope - by improving the type of glass used in the spans, the frame and the material that constitutes it;
- Passive solar systems - through, for example of the orientation and design of the spans, promoting natural ventilation and heat collection systems. These systems can be divided into passive direct, indirect, and isolated gain heating systems and passive cooling systems by natural ventilation, soil cooling, evaporative and radiative;
- Active systems - through, for example, solar systems such as photovoltaic panels or solar thermal collectors. These systems transform energy from renewable sources into energy for use.

4   Energy Rehabilitation of Residential Buildings

Nowadays, around 35% of the European building stock has more than 50 years [5]. An analysis to the energy certificates performed by BPIE (Buildings Performance Institute Europe) has concluded that around 97.5% of the building stock presents a label lower than class A [2]. However, the percentage of buildings being rehabilitated annually in Europe is low, ranging from 0.4 to 1.2% [2]. Thus, the rehabilitation of the building stock has a high energy saving potential. This section presents a discussion on the energy and economic performance of different energy rehabilitation scenarios, through a case study.

4.1 Methodology for Energy and Economic Performance

The energy performance of the case study building was assessed through a dynamic simulation using the *EnergyPlus* software [11], which reliability has been amply demonstrated [12]. To perform such simulations the energy balance method CTF (Conduction Transfer Functions) [11] was used. The CTF method allows to perform dynamic simulations with the necessary precision concerning current building solutions [13]. A life cycle approach of 30 years was considered both in energy and economic performance.

The economic performance was assessed using the methodology proposed by delegated regulation n° 244/2012 of January 16th [14], as presented on Eq. 1.

$$C_g(\tau) = C_I + \sum_j \left[\sum_{i=1}^{\tau} \left(C_{a,i_{(j)}} \times R_d(i) - V_{f,\tau}(j) \right) \right] \tag{1}$$

Where:
T: Calculation period
Cg (T): Global cost across calculation period
CI: Initial investment for measure j
Ca,I (j): Annual cost across year i for measure j
Rd (i): Annual cost across year i for measure i
Vf,T (j): Residual value of measure j in the end of the calculation period

A discount rate of 3% and an evolution of the energy costs were considered [7]. For the period between 2013 and 2030 the energy costs used were based on the EU predictions [15]. For the period between 2030 and 2046 the costs predicted in the Energy Roadmap 2050 [16] were considered. Finally, the investment costs were obtained through the CYPE price generator [17].

4.2 Case Study Building

A typical Portuguese single-family building model was used as case study building (Fig. 4). The building was considered to be located in Lisbon at an altitude of 71 m and have the following: (i) two bedrooms; (ii) a total built area of 110 m^2; (iii) an unventilated attic between the living area and the roof, (iv) an air conditioning through mobile heating and cooling systems (COP = 1; SREE = 3.5); (v) no mechanical ventilation system, thus being only naturally ventilated; (vi) three inhabitants being in there from 7 pm to 8 am in week days and all day during the weekends.

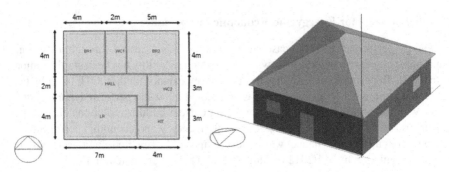

Fig. 4. Outline of the case study.

Table 1 presents the case study building solutions. These solutions were defined considering the most common building solutions used between 1960 and 1990 (period to which most of the existing buildings in the country belong) in Portugal [19].

Table 1. Case study building solutions.

Building element	Construction solutions	U (W/m² °C)
Walls	Single masonry wall with 22 cm with 2 cm of plaster both sides	1.8
Superior slab	Lightweight slab	2.8
Roof	Pitched roof with lightweight slab	3.0
Ground floor	Concrete slab covered with ceramic tile	1.7
Glazing	Single glazing (6 mm) and wooden frame	4.1

The thermal comfort temperatures recommended by the Portuguese thermal regulation [18] – 18 °C for the heating season and 25 °C for the cooling season – were considered for the analysis. The ventilation was assessed trough dynamic simulation using the *EnergyPlus AirFlowNetwork* module [11].

4.3 Rehabilitation Scenarios

Five rehabilitation scenarios were analysed (Table 2). In Scenario 1 only passive measures were considered. In Scenario 2, beyond the passive measures, more efficient building systems were defined. In Scenario 3 the measures from Scenario 1 were combined with a heat pump. Scenario 4 and 5 are the combination between Scenario 2 and 3 with a self-consumption photovoltaic kit.

Table 2. Case study building solutions.

Scenario	Description
Scenario 1	Application of 12 cm of thermal insulation (expanded polystyrene) on the externa walls (U = 0.24 W/m^2 °C) and on the superior slab (U = 0.27 W/m^2 °C), and substitution of the glazing systems for double glazing with aluminium frame (U = 2.8 W/m^2 °C)
Scenario 2	Passive measure of Scenario 1 and substitution of building acclimatization equipment for an air conditioning system (COP = 4.12; SREE = 8.53) for heating and cooling and of the DHW system by a gas condensing heater (COP = 0.97)
Scenario 3	Passive measure of Scenario 1 and substitution of building acclimatization equipment for a heat pump for heating, cooling and DHW preparation
Scenario 4	Measures from Scenario 2 with the addition of a self-consumption photovoltaic kit with a production of 1,500 W (Eren = 2,290 kWh·year)
Scenario 5	Measures from Scenario 3 with the addition of a self-consumption photovoltaic kit with a production of 1,500 W (Eren = 2,290 kWh·year)

4.4 Results

Figure 5 presents the energy consumptions obtained in each rehabilitation scenario.

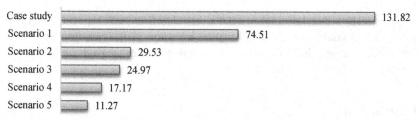

Fig. 5. Energy consumption for each rehabilitation scenario (khW/m^2·year).

It is worth to mention that the simple adoption of passive measures leads to a decrease of around 44% of the building energy consumption. These passive measures combined with more efficient but conventional building systems allow decreasing the energy consumption in 76%, approximately. The adoption of a heat pump, which is a more efficient but usually also a more expensive equipment, allows a better energy performance. However, the benefits of this system may not be sufficiently appealing to convince the user to adopt it [20]. The combination of Scenarios 2 and 3, with renewable energy production systems lead to an 87% decrease in the building energy consumption in Scenario 4 and in 91% in Scenario 5.

Table 3 presents the economic analysis regarding the implementation of the five scenarios.

Table 3. Case study building solutions.

Scenario	Initial cost (€)	Operational cost (€)	Life cycle cost (€)	Annual savings (€/year)	Payback time (years)
Ref. scenario	0	35,571.00	40,062.00	0.00	0.00
Scenario 1	9,823.00	24,692.00	36,325.00	363.00	27.00
Scenario 2	13,424.00	20,318.00	35,660.00	508.00	26.00
Scenario 3	17,171.00	24,689.00	44,415.00	363.00	47.00
Scenario 4	15,898.00	17,101.00	32,444.00	616.00	26.00
Scenario 5	19,645.00	21,472.00	41,198.00	470.00	42.00

According to the results, the payback time is too long in all five rehabilitation scenarios. Nevertheless, the scenarios with the best payback time are Scenario 2 and 4, corresponding to the adoption of passive measures, an air conditioning system, and a gas condensing heater (DHW preparation), without and with renewable energy production systems, respectively. The scenarios using a heat pump have twice the payback time than the other options. It was also noticed that regarding the payback time, the scenario based only in passive measures (Scenario 1) presents a very similar performance to the scenarios that combine such measures with more efficient energy systems (Scenario 2 and Scenario 4).

The results show that the initial cost has a very significant influence on the economic analysis of residential buildings rehabilitation measures [20]. Even in scenarios with relevant annual savings, the initial cost of the solutions makes the payback only possible in the last years of the life cycle.

5   Conclusions

Five energy rehabilitation scenarios were analysed, focusing on their energy and economic performance. This study corroborated the importance of economic analysis of rehabilitation scenarios in residential building. Here, it allowed to understand that the adoption of a heat pump may not be an interesting solution, as cheaper systems lead to a similar energy performance. However, the adoption of energy rehabilitation scenarios in residential buildings should not be seen as a way of obtaining a return of the investment. Instead, it should be a way to increase the building comfort and quality, and to decrease the building's environmental impact by using the smallest investment possible.

The energy performance of buildings in the EU is not in the overwhelming majority of cases, efficient. Its potential for improvement is significant, which can be reflected in high savings and thus, have significant contribution to a more sustainable future. In this sense, the implementation of Directive 2002/91/EC has led to a representative change in the energy dimension of buildings throughout Europe. Buildings became more energetically efficient and the population more sensitive and informed about these issues.

However, energy savings were below expectations (42%), since not all countries were able to implement the requirements entirety.

With the implementation of the EPBD-recast, in 2014, an additional 48.9 Mtoe of final energy reduction was achieved, compared to the base value of 2007, being in line with that forecast for 2020–60 to 80 Mtoe reduction of final energy [8]. Thus, with the new Directive (EU) 2018/844 reinforcements are made to the previous EPBD-recast and new targets imposed. In turn, the cost-optimal approach reveals an incentive to energy efficiency and the promotion of nZEB. However, more energy efficiency measures need to be disseminated as well as more incentive to rehabilitate.

References

1. Araújo, C., Almeida, M., Bragança, L., Barbosa, J.A.: Cost-benefit analysis method for building solutions. Appl. Energy **173**, 124–133 (2016)
2. BPIE Factsheet - 97% of buildings in the EU need to be upgraded. http://bpie.eu/wp-content/uploads/2017/12/State-of-the-building-stock-briefing_Dic6.pdf. Accessed 07 June 2018
3. BPIE Facts & Figures. http://bpie.eu/publications. Accessed 07 June 2018
4. BPIE: Europe's buildings under the microscope. BPIE, Brussels (2011)
5. EC Buildings. https://ec.europa.eu/energy/en/topics/energy-efficiency/buildings. Accessed 10 June 2018
6. EU: Directive 2002/91/EC of the European Parliament and of the Council of 16 December 2002 on the energy performance of buildings. Journal of the European Communities L 1/65 from 4.1.2003. EU, Brussels (2002)
7. EU: Directive 2010/31/EU of the European Parliament and of the Council of 19 May 2010 on the energy performance of buildings (recast). Journal of the European Communities L 153/13 from 18.6.2010. EU, Brussels (2010)
8. EU: Good practice in energy efficiency. EU, Brussels (2017)
9. Santos, J.; Sá, M.; Pereira, C.: Futureng. http://futureng.pt/reabilitacao. Accessed 10 June 2018
10. Pontes, J.: Análise multicritério de soluções construtivas para reabilitação de edifícios. Master Thesis. University of Minho, Faculty of Engineering, Guimarães (2018)
11. USDE: EnergyPlusTM Version 8.7 Documentation - Engineering Reference. US Department of Energy, U.S.A (2010)
12. Anđelković, A.S., Mujan, I., Dakić, S.: Experimental validation of an EnergyPlus model: application of a multi-storey naturally ventilated double skin façade. Energy Build. **118**, 27–36 (2016)
13. Tabares-Velasco, P.C., Christensen, C., Bianchi, M., Booten, C.: Verification and validation of EnergyPlus conduction finite difference and phase change material models for opaque wall assemblies. Technical report NREL/TP-5500-55792. NREL, Colorado, U.S.A. (2012)
14. EU: Commission Delegated Regulation (EU) no 244/2012 of 16 January 2012. Official Journal of the European Communities. https://eur-lex.europa.eu/legal-content/PT/TXT/PDF/?uri=CELEX:32012R0244&from=PT. Accessed 10 June 2019
15. EC: EU Energy Trends to 2030 – Update 2009. https://ec.europa.eu/energy/sites/ener/files/documents/trends_to_2030_update_2009.pdf. Accessed 10 June 2019
16. EC: Communication from the European Parliament, The Council, The European Economic and Social Committee of the Regions - Roadmap to a Resource Efficient Europe, COM (2011) 571 Final. UE, Brussels (2011)
17. CYPE Ingenieros, S.A.: Gerador de Preços. Portugal. http://www.geradordeprecos.info/. Accessed 07 June 2019

18. DR: Lei no58/2013 de 20 de Agosto. Diário da República, 1.a série - N.° 159 - 20 de agosto de 2013. DR, Lisbon (2013)
19. INE, & LNEC. O Parque habitacional e a sua reabilitação - análise e evolução 2001-2011. Instituto Nacional de Estatística & Laboratório Nacional de Engenharia Civil, Eds.) (2013). ISBN 978-989-25-0246-5
20. Araújo, C.: Análise da viabilidade de implementação dos princípios da construção sustentável em operações de reabilitação de edifícios residenciais portugueses. Doctoral thesis, University of Minho, Faculty of Engineering, Guimarães (2019)

Information and Technologies

Image Recognition to Improve Positioning in Smart Urban Environments

Sara Paiva[⊠], Pedro Rodrigues, and Benjamim Oliveira

Instituto Politécnico de Viana do Castelo, Viana do Castelo, Portugal
sara.paiva@estg.ipvc.pt, sara.paiva@ieee.org,
{pmiguelrodrigues,benjamimoliveira}@ipvc.pt

Abstract. This paper describes a solution and algorithm to enhance positioning in outdoor environments with high buildings to be used in a mobile application to aid visually impaired people for navigation purposes. We used an image recognition algorithm and adjusted the android app algorithm to decrease the initial error average of 85 m (without any correction from GPS obtained signal) to a 5 m error, in the final version of our solution.

Keywords: Image recognition · Outdoor positioning · Outdoor navigation · Urban mobility · Visually impaired people

1 Introduction

Obtaining a precise position is necessary in several applications we use in our day-to-day life and its improvement gains increasing importance as we witness new research and studies in these areas. The need for precise positioning can happen indoors (inside buildings) or outdoors (outside buildings). Each situation has its own particular and specific approach. We will focus on this paper in the outdoor location that is associated in the vast majority of cases to the use of GPS. In fact, most of the applications we have on our mobile phones need only this technology to guide the user on the road while driving or while walking on foot to a certain location. But when we speak of location, sometimes a margin of error of 5 m is acceptable, in other cases not. We know that the GPS position obtained has a margin of error that can be considerable, especially if we are in urban environments where there are many buildings, sometimes high, that make it impossible to receive a better GPS signal. In fact, achieving a high precision in urban environments is clearly still an issue to solve [11]. If we consider the traditional applications developed for drivers on the road, it can be considered that the current state of these applications is already advanced, considering that the effectiveness and real usefulness to end users, in this case drivers, is already quite good. However, if we look at other types of applications, such as assisting visually impaired people (VIP) people to make footpaths in the historical centers of cities, the accuracy achieved with the GPS of a commercial smartphone is still clearly insufficient. World Health Organization

S. Paiva—IEEE Senior Member.

© ICST Institute for Computer Sciences, Social Informatics and Telecommunications Engineering 2020
Published by Springer Nature Switzerland AG 2020. All Rights Reserved
P. Pereira et al. (Eds.): SC4Life 2019, LNICST 318, pp. 59–70, 2020.
https://doi.org/10.1007/978-3-030-45293-3_5

(WHO) [1] stated in 2018 that globally, it is estimated that approximately 1.3 billion people live with some form of vision impairment. The International Classification of Diseases classifies vision impairment into two groups, distance and near presenting vision impairment.

With regards to distance vision, 188.5 million people have mild vision impairment, 217 million have moderate to severe vision impairment, and 36 million people are blind. With regards to near vision, 826 million people live with a near vision impairment. Globally, the leading causes of vision impairment are uncorrected refractive errors and cataracts. According to the study published in 2017 by The Lancet Global Health, compared to 1990, the number of blind people increased by 17.6% and that of individuals with moderate and severe problems about 35.5%. The same study also predicts that by 2020 there will be 237.1 million people with moderate to severe vision impairment and the number of blind people will be 38.5 million. By 2050, these figures are estimated to be about 587.6 million and 114.6 million individuals [14]. In Portugal, according to the 2011 Census, there are 900,000 cases of visual impairment, of which 28,000 are estimated to be entirely blind [2]. Approximately 80% of all vision impairment globally is considered avoidable. The majority of people with vision impairment are over the age of 50 years. A person's experience of vision impairment varies depending upon many different factors. This includes for example, the availability of prevention and treatment interventions, access to vision rehabilitation (including assistive products such as glasses or white canes), and whether the person experiences problems within accessible buildings, transport and information. According to this reality, technological research related to the promotion of accessibility and quality of life of the VIP has been under study for several years. Several solutions have also been developed sustained by the evolution of computational processing and data processing. Technologies have evolved and reached a big level of maturity that allow new technologies, approaches and algorithms to solve problems related to VIP more effectively, always with the goal of making their lives less affected by this deficiency to the greatest extent possible. In this article, we present a practical case study in the city of Viana do Castelo, Portugal, where we worked alongside an association of blind and partially sighted people (with whom previous work has already been done to help them through computer solutions in bus navigation in the center as well as pedestrian navigation [7, 10]). In this paper, we present an algorithm we developed and that is incorporated in a mobile app to make image recognition in order to allow the correction of a received GPS position. Images are captured from the visually impaired people (VIP) smartphone camera which triggers the algorithms to identify that image with a pre-loaded geo-referenced image in a dataset, so the GPS position of the user can be corrected and enhanced compared to the position directly obtained from the GPS.

The rest of the paper is structured as follows: the next section presents a literature review on positioning technologies. Section 3 explains the methodology followed for this project and the next chapter has all details about its implementation. Section 5 presents results and discussion and Sect. 6 presents main conclusions.

2   Positioning Technologies

GPS (Global Positioning system is probably the major outdoor positioning technology we have at the current moment, assisting several location-based services. It is used in industries such as aviation, nautical navigation, and land surveying, to personal and commercial applications such as driving navigation, people tracking, location sharing [21]; also, it can be of great help and fundamental to make a difference in the lives of people with disabilities facilitating day-to-day tasks. We have referred already to the imprecision of GPS position, namely in places with high-building, that can go from 1 to 20 m. Considering this, one study to overcome this situation is to use GPS to implement a high-accuracy global positioning solution and human mobility captured by mobile phones [22]. Authors developed a prototype system, named GloCal, and conduct comprehensive experiments in both crowded urban and spacious suburban areas. The evaluation results show that GloCal can achieve 30% improvement on average error with respect to GPS. Another study combines GPS with smartphone-enabled dead reckoning to produce a high-accuracy global positioning solution based on GPS and human mobility captured by mobile phones [21]. The performance comparison of four self-position fixing methods is made in [8] where authors analyze RSS Statistics, Least squares, Weighted LS (WLS) and Constrained WLS. Other authors provide a solution that makes use of a handheld device [3]. Their ideas to estimate the 3D location and 3D orientation of the phone camera based on the knowledge of the street objects. The use of vision process to generate a3D model with inertial and magnetic measurements is yet another proposal to update a Pedestrian Dead-Reckoning process and to improve the positioning ac-curacy [4]. Other contribution investigates the spatial correlation of multi-path error to estimate the multipath error at a pedestrian by using a regression model and leveraging the multipath errors at nearby points [12]. In [20] authors propose a new positioning algorithm based on signals only with pseudo range error modeling in association with an adapted filtering process. Another approach corrects signals using the European Geostationary Navigation Overlay Service (EGNOS) [19] and also presents an overview of the accuracies that can be achieved with different modes of GPS positioning. In [18] a method to mitigate multiple path in GNSS applications is proposed. Their principle is based on the addition, to the geographic map of a city or a neighborhood, of a supplementary information that consists of the correction of error caused by the multiple paths. Issues and problems associated with GNSS are presented and discussed in [13]. GPS and GLONASS were chosen for a comparison on accuracy and precision. The tightly-coupled single-frequency multi-GNSS RTK/MEMS-IMU integration is proposed in [9] to provide precise and continuous positioning solutions in urban environments. In a similar way, authors in [23] describes and applies a DGNSS-correction projection method to a commercial smartphone. An example of the use of images to enhance positioning is used in [15] where authors use the camera and a GPS-tagged image database to identify with precision the position the user stands. In [6] authors propose to improve the localization of scene items based on state-of-the-art map data, combined with a coarse and cheap position estimation as provided by standard GNSS. In [16] authors propose a method to geo-locate a mobile device by recognizing what is captured by its camera. A visual recognition algorithm in the cloud is used to identify geolocated reference images that match the cameras view. Some case studies are also presented and found in the literature. A study on the performance of BeiDou Navigation Satellite System in urban environments have been made [5] in Wuhan, central China. The compared data (visibility, position dilution of precision and the positioning accuracy) showed that the single point positioning accuracy of the GPS/Beidou dual system is much better and more stable than single BeiDou system. The POI Explorer mobile application aims to help VIP in their spatial orientation and urban navigation, using a smartphone, its sensors and data connectivity. Interaction is made through text-to-speech and notifications [17].

3   Methodology

The methodology followed for this project consisted in the concrete identification of the type of signal collected in the historical center of the city of Viana do Castelo, in one of its busiest streets. The goal was to realize the real mistake and therefore the impossibility of using only GPS to develop an application that allows VIP to move around the city and get relevant information about their location that allows them to gain more autonomy. After that, we designed an algorithm that compares captured camera image from the commercial smartphone to compare it to a preloaded geo-referenced image in a dataset so position can be enhanced compared to GPS received position. We made first round of tests that showed necessary improvements to be incorporated in the algorithm which we did to create a second version of it. Final results will be presented based on this second version of the algorithm and tests made in the field.

4   Implementation

Stage 1 - GPS Precision Analysis
As aforementioned, the first step was to pick a strategic street in the historical center and gather some data on the reliability and precision obtained. The street is shown in

Fig. 1. Street chosen for the tests made

Fig. 1 and it is possible to see it has high buildings which naturally affects received GPS signal.

The tests made included going down from one side and then going up from the other side, covering both sides of the beginning of the street. 167 positions were chosen, the exact position was marked in a map using a mobile app and the received position was compared to it. The average error received, in meters, was of 85.5. The minimum error obtained was 6.3 m and the maximum 205.1 m. Figure 2 shows the exact location of tester going down the road (green markers) and the obtained position through GPS (red markers). It is possible to see that most locations were received has been on the other side of the road, which makes it completely impossible to use only GPS to create a location-based app to aid VIP in their navigation needs.

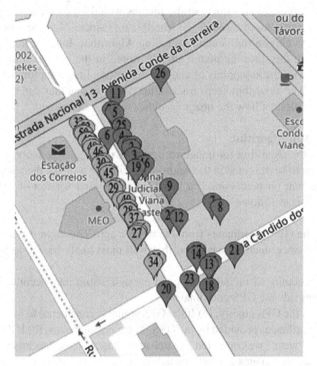

Fig. 2. Correct position vs. received GPS position without any correction (Color figure online)

Stage 2 – First Algorithm
This stage consisted in developing the algorithm that would allow for an image from the VIP cellphone camera to be compared with a dataset of geo-referenced images, so a better position than the GPS could be returned to a mobile app to be used for navigational purposes to aid VIP. For building the dataset we captured some relevant objects throughout the street that could serve as a reference point and allow for the image

recognition to be successful. At this point, we were aware that the success of the entire algorithm was mostly related to the performance of the image recognition algorithm.

Choosing the Image Recognition Algorithm

OpenCV (Open Source Computer Vision Library) is a well-known open source computer vision and machine learning software library, with different comparison methods. The Histogram Comparison Method is the fastest and simplest to adapt regarding image comparison. After we have analyzed this algorithm, we concluded it would not be appropriate to use in this case, because images are compared mostly by color, and as there are several buildings with the same color, it could lead to a huge margin of error, since two images containing buildings of the same color would be considered the same. The Template Matching method basis its comparison mainly depending of its size, which would also not work in this project, since all the images would be considered equal, assuming they would be captured through the smartphone camera. A more thorough search has allowed us to choose the Feature Matching Algorithm. Initially the key points of a given image (in this case the image already stored in the database) are chosen and later compared with the key points of another image (the image captured by the smartphone camera). The algorithm keeps track of the points so that they can be detected and recognized regardless of how the image is, either re-scaled or turned.

The Android App Algorithm

After choosing the algorithm for image recognition, it was time to incorporate it in the global algorithm to be executed in the Android mobile app to make the tests to conclude about enhancement on precision. The behavior of the first version of the algorithm, shown in Fig. 3, is as follows:

- We receive the GPS coordinates from the mobile app. These are the coordinates we will try to enhance since, as we know, they will most likely be wrong by a few tens of meters;
- Check if the quality of the signal is good enough, using the "accuracy" parameter received alongside the GPS coordinates;
- Depending on the GPS quality: (a) If the GPS signal is considered to be good enough, we use the coordinates provided by the GPS. The algorithm ends. (b) If the GPS signal is considered "weak" we computed a search area (based on the "accuracy" parameter) and retrieve all the images geo-referenced inside that same area;
- Two parallel actions are made at this point: (a) Obtain images on the range of the search area. Each image on our dataset is geo-referenced. Using the area previously calculated we retrieve all the images that are inside it. (b) Obtain image from the smartphone camera;
- Compare captured image with images from the dataset. To find out which image matches best the image retrieved by the smartphone we have to loop through all of the images we have filtered;
- Retrieve position of the most similar image. Once the previous loop ends, we will retrieve the image with more similarities to the image being captured by the smartphone camera;

- Place a marker on the map in the new corrected position and use it to inform the user of his location or nearby obstacles.

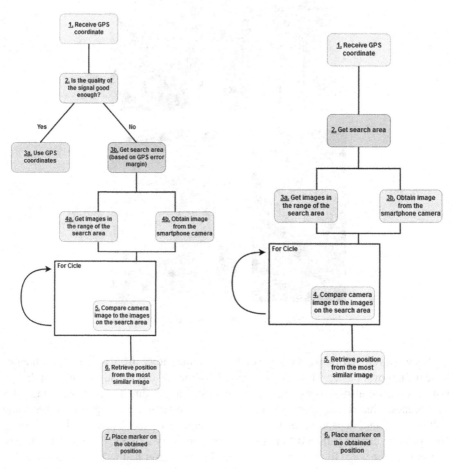

Fig. 3. Algorithm - version 1   **Fig. 4.** Algorithm - version 2

When testing on the field this first version, two main problems were noticed. The first has to do with the "accuracy" parameter returned by the GPS. It cannot be used to decide about precision of the signal as plenty of signals are marked as having a good accuracy with tens of meters of errors. The second change has to do with how the dataset was being created. Figure 5 shows some images in the dataset, with which captured images on the smartphone were to be compared with. The image recognition algorithm did not have a good behavior as its strength is to detect recognizable symbols which are small in these first images and therefore, we had few matches.

Fig. 5. Examples of images from the first dataset

So, we developed a second version of the algorithm shown in Fig. 4. The main changes were the removal of step 2 regarding accuracy, for the reasons already mentioned. In this last version, we also narrowed dataset number of images directly after receiving GPS coordinates. This is an important and necessary step as image recognition takes some milliseconds and, depending on the number of images on the dataset, and considering GPS position can be gathered from the smartphone every 2 or 3 s, the algorithm had to be ran in a fast way. To achieve this, images in the dataset had to have a considerable small size, otherwise several seconds would be needed to run the algorithm which would not suit the mobile application itself.

5   Results and Discussion

As previously explained, this project consisted of three main phases. The first one ana-lyzed the accuracy of the signal obtained without any correction. In a second phase, a dataset of images was defined and developed a first algorithm. Finally, in phase 3, an improved version of the algorithm was developed, according to tests made in phase 2. In each phase, our purpose was to measure the accuracy of the signal received compared to the real location of the user. In the first phase we obtained an average of 85 m of error comparing the real position with the obtained GPS position. In the second phase, we used the first version of the algorithm and the average error was of 54 m. At this point, we expected to have better results and understood the result was very much originated by the incorrect match/recognition of the image the smartphone was capturing, comparing to the ones on the dataset. At this point, we adjusted the algorithm and also changed the dataset of images, so they are mainly composed of brands/logos or very relevant and significant sub-parts of the image being captured by a camera. Figure 6 shows some images used in the second dataset which are significantly different from the ones used in the first dataset shown in Fig. 5. This allowed the image recognition algorithm to have a much better behavior and recognize images more successfully.

Fig. 6. Examples of images from the second dataset

With the changes made in the second algorithm, which corresponds to phase 3 of our process, the obtained average error went down to 5 m, as can be seen in Fig. 7.

Similar to Fig. 2, we have measured correct position versus position obtained with version 2 of the algorithm and results are, as can be seen in Fig. 8, much more accurate.

The improvement was significant (reduction of approximately 80 m in average compared to phase 1 and reduction of approximately 50 m compared to phase 2), but we know it can be even better and we need it to be we this algorithm can be used in mobile solutions for visually impaired people. Images are correctly detected, and position is corrected but 2 or 3 m ahead, the recognition of the image is still done, and corrected position keeps being the same of the coordinates in the image from the dataset which contributes to some errors. To further decrease the obtained error number and solve this problem, we need also to use smartphone sensors, so we detect how much the user has walked since the first correction considering a given image and introduce an additional correction, that not only is based on the coordinate of the image that made a match but also on the user's movement.

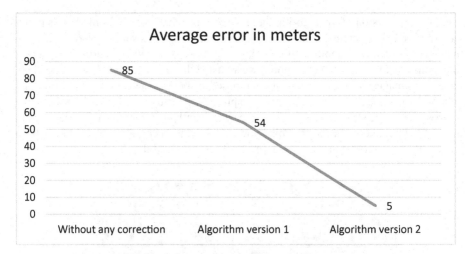

Fig. 7. Comparison of average meters obtained in each of the 3 phases

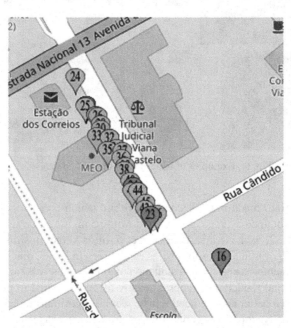

Fig. 8. Correct position vs. received GPS position using algorithm version 2 (Color figure online)

6 Conclusions

In this paper we have presented an on-going work that has the main goal to develop an algorithm to achieve better precision in positioning that can be used in a mobile application to aid visually impaired people to walk in a more autonomous way within cities. For this, we used image recognition to correct the GPS position received by the satellite that has, as we know and if used alone, still a big error. Our approach started by choosing a street with much usage in the city of Viana do Castelo and understand precision of GPS received signal. We found out average error is approximately 85 m, considering that are high buildings. So, we implemented the first version of an android app algorithm that makes use of an image recognition algorithm so the captured image from the smartphone camera is compared to a preloaded dataset of geo-referenced images. This first approach allowed us approximately to a 54 average error meter. It was still a big number mainly because of the images in the dataset were not the most appropriate as they captured a large part of the buildings. We had to change to significant markers and symbols only so recognition could be more successful. The second version of the algorithm achieved a 5 m error average which was a significant increase. To obtain further results we will, in a next stage, combine work done so far with smartphone sensors to allow to narrow down the error m obtained.

References

1. Blindness and vision impairment (2018). https://www.who.int/news-room/fact-sheets/detail/blindness-and-visual-impairment. Accessed 13 June 2019
2. Associação dos cegos e amblíopes de Portugal (2019). http://www.acapo.pt/deficiencia-visual/perguntas-e-respostas/deficiencia-visual. Accessed 13 June 2019
3. Antigny, N., Servires, M., Renaudin, V.: Hybrid visual and inertial position and orientation estimation based on known urban 3D models. In: 2016 International Conference on Indoor Positioning and Indoor Navigation (IPIN), pp. 1–8 (2016)
4. Antigny, N., Servires, M., Renaudin, V.: Pedestrian track estimation with hand-held monocular camera and inertial-magnetic sensor for urban augmented reality. In: 2017 International Conference on Indoor Positioning and Indoor Navigation (IPIN), pp. 1–8 (2017)
5. Xiang, X., Bian, S.: Research on BeiDou positioning performance in urban environments. In: 2013 IEEE International Conference on Signal Processing, Communication and Computing (ICSPCC 2013), pp. 1–5 (2013)
6. Cao, G., Damerow, F., Flade, B., Helmling, M., Eggert, J.: Camera to map alignment for accurate low-cost lane-level scene interpretation (2016)
7. Faria, P., Curralo, A.F., Paiva, S.: A case study for the promotion of urban mobility for visually impaired people. In: Paiva, S. (ed.) Mobile Solutions and Their Usefulness in Everyday Life. EICC, pp. 65–82. Springer, Cham (2019). https://doi.org/10.1007/978-3-319-93491-4_4
8. Krishna Reddy, D., Achanta, D., Satya, V.: Mobile position estimation with RSS based techniques in an urban city with multiple multi-storied structures, pp. 234–234 (2014)
9. Li, T., Zhang, H., Gao, Z., Chen, Q., Niu, X.: High-accuracy positioning in urban environments using single-frequency multi-GNSS RTK/MEMS-IMU integration. Remote Sensing **2018**, 205 (2018)
10. Lima, A., Mendes, D., Paiva, S.: Mobile solutions for visually impaired people: case study in Viana do Castelo Historical Center. In: 2017 12th Iberian Conference on Information Systems and Technologies (CISTI), pp. 1–6 (2017)
11. Lin, H., Ye, L., Wang, Y.: UWB, multi-sensors and wifi-mesh based precision positioning for urban rail traffic. In: 2010 Ubiquitous Positioning Indoor Navigation and Location Based Service, pp. 1–8 (2010)
12. Patou, Y., Obana, S., Tang, S.: Improvement of pedestrian positioning precision by using spatial correlation of multipath error. In: 2018 IEEE International Conference on Vehicular Electronics and Safety (ICVES), pp. 1–8 (2018)

13. Przestrzelski, P., Bakua, M.: Study of differential code GPS/GLONASS positioning. Ann. Navig. **21**, 117–132 (2014)
14. Bourne, R.A., et al.: Magnitude, temporal trends, and projections of the global prevalence of blindness and distance and near vision impairment: a systematic review and meta-analysis. The Lancet Global Health **5**, e888–e897 (2017)
15. Salarian, M., Manavella, A., Ansari, R.: Accurate localization in dense urban area using google street view image, pp. 485–490 (2015)
16. Sanjuan, D.M., Adamek, T., Bonnin, A., Trzciski, T.: Enhancing global positioning by image recognition (2011)
17. Skulimowski, P., Korbel, P., Wawrzyniak, P.: POI explorer a sonified mobile application aiding the visually impaired in urban navigation, pp. 969–976 (2014)
18. Titouni, S., Rouabah, K., Attia, S., Flissi, M., Khababa, O.: GNSS multipath reduction using GPS and DGPS in the real case. Positioning **08**, 47–56 (2017)
19. Verbree, E., Tiberius, C., Vosselman, G.: Combined GPS-galileo positioning for location-based services in urban environment. LBS TeleCartography **66**, 99–107 (2003)
20. Viandier, N., Nahimana, D.F., Marais, J., Duflos, E.: GNSS performance enhancement in urban environment based on pseudo-range error model. In: 2008IEEE/ION Position, Location and Navigation Symposium, pp. 377–382 (2008)
21. Wu, C., Yang, Z., Xu, Y., Zhao, Y., Liu, Y.: Human mobility enhances global positioning accuracy for mobile phone localization. IEEE Trans. Parallel Distrib. Syst. **26**(1), 131–141 (2015)
22. Wu, C., Yang, Z., Zhao, Y., Liu, Y.: Footprints elicit the truth: improving global positioning accuracy via local mobility. In: 2013 Proceedings IEEE INFOCOM, pp. 490–494 (2013)
23. Yoon, D., Kee, C., Seo, J., Park, B.: Position accuracy improvement by implementing the DGNSS-CP algorithm in smartphones. Sensors **16**, 910 (2016)

An Hybrid Novel Layered Architecture and Case Study: IoT for Smart Agriculture and Smart LiveStock

Pelagie Houngue[1], Romaric Sagbo[1(✉)], and Colombiano Kedowide[2]

[1] Institut de Mathématiques et de Sciences Physiques, Dangbo, Benin
{pelagie.houngue,romaric.sagbo}@imsp-uac.org
[2] Université TELUQ, Québec, Canada
colombiano.kedowide@gmail.com

Abstract. The meteoric rise in the number of connected objects in our daily lives is proof that data transmission and improvement of services related to our activities are a permanent and urgent concern. Communicating objects transform our behaviors, our habits and our society in general. Despite the significant progress in the field of Internet of Things (IoT), much remains to be done especially in developing countries. In the field of e-agriculture, digital production techniques are not enough to guarantee a better yield and safeguard crops. Thus, in this work, we have focused on the resolution of the problems related to transhumance, given the expansion of the phenomenon in developing countries. Indeed, during transhumance, passages intended for animals, called corridors may not be followed by the breeders. This can lead to deadly clashes between herders and farmers. Our vision is to help through the implementation of a smart guidance system based on IoT Technologies, herders to better control their livestock following the predefined corridors from north to south of Benin and vice versa. In order to help farmers to save their crops in case of flood, our system will integrate a prediction module that will enable them to anticipate natural events such as flooding in the heavy rainy season. In this paper, our researches will therefore focus on the proposal of a multi-level architecture that can enable us to achieve the aforementioned objectives.

Keywords: IoT · Smart · Prototype · Sensors · E-agriculture · Smart-village · Livestock · Farmer · Big data

1 Introduction

One of the latest breakthroughs in the Internet world was the Internet of Things (IoT). The term IoT dates back to 1999 [1] and is a science that allows multiple objects to communicate with each other and/or with the cloud. This technology currently promises to positively and dramatically change human's life through

P. Pereira et al. (Eds.): SC4Life 2019, LNICST 318, pp. 71–82, 2020.
https://doi.org/10.1007/978-3-030-45293-3_6

these connected objects. Connected objects rely on embedded chips and sensors that connect them to the Internet and give them new features. Thanks to IoT, it is also possible to facilitate the convergence between several disciplines such as embedded systems, Artificial Intelligence, Big Data, Open Data, Cloud computing, Machine-to-Machine communication (M2M), etc. Once upon a time, isolated [2]. Gartner Institute predicts that nearly 21 billion connected devices could be in circulation by the end of 2020, two-thirds for private uses and the rest for the professional world. The Internet of Things has become today, a required tool for the rapid expansion of several sectors such as health, agriculture, transport, livestock, manufacturing, logistics, building, home automation, etc.

The fields of agriculture and livestock are of particular interest to us in this paper, as agriculture and livestock are the keys to economic development and poverty reduction in developing countries. For a long time and for a variety of reasons, including structural and technological constraints, misguided national policies and an unfavorable external economic environment, the potential of these two sectors has been ignored. The implementation of an intelligent farming system makes it possible to cope effectively with both productivity and low income issues and those related to climate change. The Food and Agriculture Organization (FAO), a specialized agency of United Nations, estimates that food production must increase by at least 60% to meet the demand of the 9 billion people expected to populate the world in 2050. It states that one people between eight is currently food insecure [3]. It is therefore urgent to find sustainable solutions to ensure global food security in the coming decades. As for the livestock sector, the use of technologies makes it possible, for example, to monitor animals, to know in real-time the state of their health and well-being, to acquire information on livestock, to improve the management of herds and to better guide them during transhumance through predefined corridors [4], etc.

The remaining of this paper is organized as follows. Section 2 presents the context and the motivation. Section 3 describes our multi-layered IoT architecture. Finally, Sect. 4 presents the related work and Sect. 5 concludes our work.

2   Context and Motivation

Transhumance is a phenomenon that involves migrating livestock from one place to another by traveling several kilometers and crossing fields of crops. The goal is to allow livestock to feed properly and reproduce. Generally, the non-respect of the passages planned for the animals called corridors can lead to a destruction of the harvests and lead to deadly clashes. This can have a negative impact on the development and economy of a country, especially in developing countries. In addition, it is noticed that the phenomenon of flood, can cause very serious damages and have significant impacts on the quality of life of the people living in rural areas. The damages can be both material and human. These two factors led us to think of the establishment in Benin, particularly in the Ouémé valley (second largest valley in the world) [5], of a project that we named THIN-GALIVE (**TH**e Internet of thi**N**gs for intelli**G**ent **A**griculture and **L**ivestock

In the Valley of ouEme) that aims to introduce the Internet of Things [1] in agriculture and livestock, including the transhumance of animals.

This work will allow us, in addition to getting the usual information on a given crop field (soil condition, growth level, humidity, temperature), to provide reliable information about the imminent arrival of a flood so that the farmers can anticipate and save the harvest by sheltering more crops and also to reduce the risks of frequent drowning.

On the other hand, researches conducted in the course of this project will enable pastoralists in transhumance to effectively monitor livestock by following the corridor defined for their passage from the South of Benin and reduce clashes with farmers. Those clashes include the slaugther of cattle, the destruction of the harvest and the loss of human life. So, breeders will be able to have real-time information if they deviate from the corridors and have frequent information on the state of the grazing along the corridor.

THINGALIVE project will therefore lead to a more intelligent agriculture that will allow the farmer to be fulfilled and better enjoy the fruits of his toils. On the other hand, farmers will benefit from a transhumance that is both peaceful and intelligent in the predefined corridor by also providing useful information on the areas of the corridor where the pasture is well provided throughout the displacement or position of the water bodies.

Thus, THINGALIVE project will allow the use of technologies related to the IoT to achieve the following objectives:

- Better inform farmers and breeders to ensure peaceful relationships
- Make profitable the activities of farmers and breeders for a better development impact
- Reduce or eradicate the deadly clashes between farmers and breeders
- Create smart villages that are better organized to face the flood and avoid the destruction of the crops
- Preserve the security and privacy of the information.

The THINGALIVE project is still in its earlier stage and is divided into several phases, given its scale. Thus, this paper will only focus on the proposal of the architecture that will gradually lead to the achievement of the objectives mentioned above.

Thus, the proposed architecture is intended to allow:

- breeders to keep their herds as much as possible during transhumance in the corridors provided for this purpose.
- farmers to intervene in time to prevent the total destruction of their crops in case of negligence on the part of a breeder.
- the collection, storage and processing of data necessary for rapid decision-making.

3 High Level IoT Architecture for THINGALIVE

This section presents a short overview on IoT reference model, describes the architecture proposed for THINGALIVE project and compares the two architectures to show how our architecture fulfils the IoT basic requirements.

3.1 Reference IoT Model

Due to the rapid growth of the IoT, to cope with the lack of standardization of IoT model, it is important to have a reference architecture. This model needs to give a common view of the different levels of devices, tools, standards and protocols needed to make IoT running from the data generation by the devices, their processing and transmission, to their use by applications. Many models and architectures have been proposed such as Lamda Architecture [6,7], Kappa Architecture [7], IoT World Forum reference model [2] and many others, depending of the covered domains. The authors of [2] proposed a multilevel model to give clear definitions and descriptions applicable to elements and functions of IoT systems and applications. Then, the reference model has been proposed to bring simplification, clarification, identification, standardization and organization into IoT system setting up. The reference model has seven levels which are shown in Fig. 1:

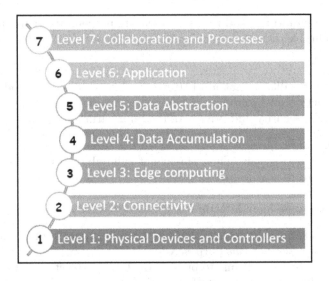

Fig. 1. IoT reference model

- Level 1: Physical Devices and Controllers. This level is composed by physical devices and controllers that might control many devices/endpoints which send and receive information and can give sometimes a first level of processing.

- Level 2: Connectivity. This level gives a reliable, timely information transmission between devices across networks on the one hand and between the network and the available functions at Level 3 on the other hand.
- Level 3: Edge Computing. This level ensures the conversion of the network data flows into information in order to serve the next level for storage and high level processing.
- Level 4: Data Accumulation. This level captures network data in movement from the previous levels and stores them, using different database technologies for non-real-time applications.
- Level 5: Data Abstraction. With IoT, data originate from many sources. This level aims at focusing on presenting data and its storage in such a way to enable easier implementation and provide powerful applications. This level processes the data to make them ready for the next level. Data protection with appropriate authentication and authorization is also provided.
- Level 6: Application. This level provides information interpretation. Different types of application can be developed depending of the nature of the data acquisition devices and business needs from data available at Level 5. This will allow getting better end-user experience for mobile users, business analysts and for monitoring purposes.
- Level 7: Collaboration and Processes. This level highlights the fact that many people need to collaborate to make working an overall IoT ecosystem through many processes that involve many data generated by IoT systems.

3.2 High Level Multi-layered Architecture for THINGALIVE

This section presents our proposed multi-layered architecture for smart agriculture and livestock based on IoT technologies.

We identified three main requirements for our architecture which are :

- Completeness: our architecture should meet the key properties of IoT reference model architecture.
- Flexibility: this gives our architecture the possibility to be independent from any technology and protocol constraints.
- Simplicity: this ensures our architecture to be clear and easy to understand.

Figure 2 shows the multi-layered architecture proposed for the THINGALIVE project. Our architecture is firstly composed of a stack of five main layers that highlight the required functionalities of an IoT-based architecture. Those layers are: Data Acquisition Layer, Connection Systems/Technologies Layer, Data Processing and Storage Layer, Service Layer and Notifications Layer.

Moreover, it integrates a lateral stack of two additional layers that show interaction with the other layers for monitoring and give access to the services provided to the different actors. Those are: Monitoring System and Stakeholders layers.

Following is the description of the proposed architecture.

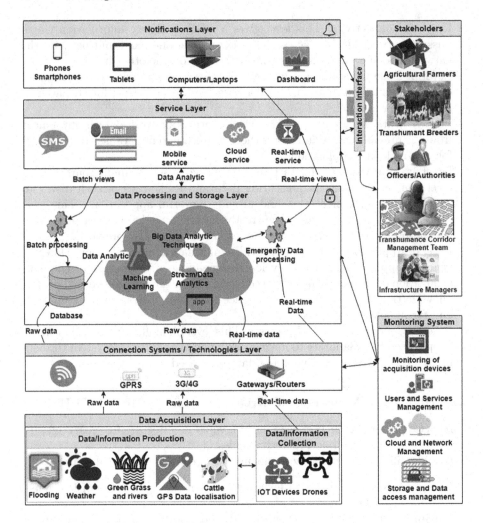

Fig. 2. THINGALIVE multi-layered IoT architecture

- **Data Acquisition Layer**

  This layer is composed of elements that produce information collected by IoT devices, sensors and actuators. The data can be produced by weather, flooding situation, the position of the green grass, rivers and cattle. Drones and IoT devices are used to collect data that are sent through the next layer using different communication technologies. The data transmitted can be processed in real-time or with delay depending of the usage.

- **Connection Systems/Technologies Layer**

  This layer provides communication functions for IoT devices. It allows the transmission of the data flows from the sensors to the next level for storage, processing and use. There are many proprietary technologies and protocols in use, but the standardization increases quickly and allows more interoperability. IoT uses several well-known and emerging technologies, namely Wi-Fi, Bluetooth, ZigBee, Z-Wave, 6LowPAN, Tread, Neul, LoRaWAN, Near Field Communication (NFC), Sigfox, etc., in addition to 2G/3G/4G cellular technologies [2,8,9]. Depending of the purpose of the collected data, they could be sent either to the database for future use or directly to application for real-time exploitation. This layer is designed to be reliable and is able to deliver information in real-time between devices over the network by implementing various communication protocols. Both protocols in common use on the standard network and IoT-aware protocols will be used. For example, LoRaWAN or SigFox [8] technology can be used to transmit data from sensors to the place they will be analyzed in order to act and take decisions.

Among the common protocols available on TCP/IP stack, HTTP is used for some IoT devices. However, it does not support IoT devices environment constraints, such as low memory, low bandwidth, low power and others. Moreover, many web-derived protocols are also proposed but they have some limitations. In the literature, more suitable IoT protocols have been proposed to give better facilities for communication in IoT world. The most popular protocols are CoAP (Constrained Application Protocol) [10] and MQTT (Message Queuing Telemetry Transport) [11,12]. Depending of the IoT devices and the usage, our architecture will use one that suits a case.

– **Data Processing and Storage Layer**

This layer is the most important of our architecture, since it provides data processing and storage. Data processing is performed according to the target of utilization by converting network data flows into information. This layer tackles high-volume data analysis and transformation since IoT devices may generate a lot of samples of data that need to be analyzed in order to take decision and to be stored. The analysis can be performed in real-time or delayed based on stored data. The existing analysis techniques and tools are based on common techniques such as machine learning, neural networks [13,14] and big data analytic techniques [15] used in cloud computing to process largest data. Emergency data are processed directly and sent to related service to serve for quick handling of events. Batch processing is based on stored raw data used to build analytic solution in order to act and perform prediction. This layer provides understandable information to the higher layer. Data are filtering, cleanup and aggregate for utilization by each service using data in motion and/or data at rest. For storage, streaming or continuous data may be stored on big data system such as Hadoop [16], MongoDB [17] and non-real-time data may be stored in relational database where data are always available. Real-time data is usually used to setup alert systems. This layer provides functionalities such as multiple data formats reconciliation, consistent semantics of data assurance across sources, transformation of data for higher level application, consolidation of data into one place using ETL (Extract, Transform and Load) or data replication, access to multiple data stored through data virtualization and data protection. Today, OLTP databases [18,19] and data warehouses are used to carry and manage the amount of data generated by multiple IoT devices.

- **Service Layer**

  This layer is the one that helps to make use of the amount of data provided by the IoT systems. The interpretation of information is performed by this layer from real-time data and/or non-real-time data. The applications developed on the stack of this layer are based on different requirements, such as monitoring, prediction and control. The data transformed from the previous layer are used here to provide various functionalities. For example, alert system can be setup using real-time data to provide quick response time in case of disaster. Furthermore, analytic applications is setup to interpret data for business decisions. This layer is the one that provides interactions between the different actors and the IoT systems setup for smart Agriculture. The services developed by our architecture can be based on restful web service and many protocols designed for IoT.

- **Notifications Layer**

  This layer provides interaction with users through their devices. The services provided by the IoT systems are reachable from the common communication devices (laptops, phones, tablets, etc.). Web services using SOAP or REST and restful API are used to provide services to the end user. Taking into account user's position, the information received will provide him with the possibility to make a decision. This layer provides a dashboard to follow different events from the devices and to manage the different devices available in the IoT system.

- **Monitoring System**

  The first additional layer provides an interface to manage the devices, the tools, the services and our architecture end-users. This layer can take into account the monitoring of the IoT devices, the users and services management, the cloud infrastructure and services management, the communication network management and the databases administration.

- **Stakeholders**

  The use of the services developed around the data generated by the devices allows many interaction between the people involved and the processes that will run behind the real-time or non-real-time data that move through the network.

  The second additional layer is composed of the main actors of our architecture. Five of them are formally identified. They are:

  - The Agricultural farmers. This actor is the farmer that has the land and performs agricultural activities on it.
  - The Transhumant breeders. This actor is the breeder who travels along the corridors to have green grass and rivers for the cows and other animals.
  - The Officers and Authorities. The officer or the authority is the actor that can be called or can receive a message notifying that something wrong is happening so that the security company tries to solve the issue.
  - The Transhumance Corridor Management Team. This team is composed by many people, from the government, geographers to the representatives

of farmers and breeders. This committee defines each year the corridor that will be followed by the breeders.

- The infrastructure managers. The infrastructure will be managed by our team and may be used by some farmers that will be equipped with our IoT devices and materials. Some breeders also may have the devices placed on some cows to visualize in real-time their position along the corridor in order to send an alert when they didn't follow the corridor.

Our architecture provides six main functions that are spread over the seven layers namely: Collection, Storage, Analysis, Prediction, Adaptation and Output (Application).

3.3 THINGALIVE Architecture vs IoT World Forum Reference Model

This section compares the IoT World Forum reference model [2] with our proposed architecture in order to show how our architecture is complete by taking into account all of the functionalities delivered by the seven layers of the reference model. Our architecture has two layers less than the reference model and aggregates many functions on each layer to reduce the complexity. Figure 3 shows a comparison between our architecture and the IoT World Forum reference model. The links from the reference model to our architecture show the correspondence between the layers on the both sides.

Our Data Acquisition Layer provides the functionalities delivered by the first layer of the reference model, Physical Devices and Controllers.

Our Connection Systems/Technologies Layer plays the role of the second layer which is Connectivity.

Our third layer, Data Processing and Storage Layer is the most important and aggregates the functionalities of the third, fourth and fifth layers of the

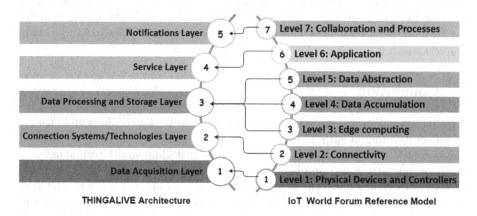

Fig. 3. Comparison between THINGALIVE architecture and IoT world reference model

reference model, Edge Computing, Data Accumulation and Data Abstraction. This layer is the core of our architecture.

The fourth proposed layer, named Service Layer tackles the functions given by the application layer of the reference model. The additional Monitoring System layer can also be associated for services delivery one.

The last layer named Notifications layer proposed the role played by the seventh layer of the reference model, Collaboration and Processes. The additional Stakeholders and Monitoring system layers can be associated to add more functionalities.

This comparison shows that our architecture fulfils the basic requirements of IoT systems.

4   Related Work

The Lambda Architecture presented in [6] is a robust system for massive amount of data collection and processing, but its core is centralized in the data centers or in the cloud. This architecture can limit the effectiveness of the analytics to quickly respond.

The authors of [2] proposed a layered IoT architectural reference model that provides a clean, simplified perspective on IoT and solutions that make easy the data collection, processing and service. This proposition splits the IoT problem into smaller parts and provides interoperability between devices in addition to security. This architecture needs to be adapted according to each application, in particular in african context where few communication technologies are available.

The authors of [20, 21] show that the problem between breeders and farmers is real, but they didn't provide any concrete IT solutions to overcome the situation.

The work in [4], proposed a solution that addresses the mapping of the corridors without showing a full applicable IT solution to reduce clashes between breeders and farmers.

Many existing IoT-based solutions [22–24] are partial whereas the solution we proposed in this paper, aims to provide an integrated architecture that is useful for both breeders, farmers and authorities. This architecture tries to handle many aspects which are common for breeders and farmers.

5   Conclusion

The IoT agricultural applications are making it possible for ranchers and farmers to collect meaningful data. Large landowners and small farmers must understand the potential of IoT market for agriculture by installing smart technologies to increase competitiveness and sustainability in their productions. With the rapid growth of the population, the demand can be successfully met if the ranchers, as well as small farmers, use agricultural IoT-based solutions in a prosperous manner. This is the reason why we wanted to propose an architecture whose main objectives were to allow both farmers and breeders to be alerted in case of deviation of herds during transhumance. This is only possible through the data collection, storage and processing system that can assist in reliable decision-making inherent in good herd management. The architectural solution proposed in this paper represents an important step that should create the hope of definitively solving the conflicts related to the destruction of crops due to the transhumance of animals on the one hand and issues related to crops destruction due to flooding occurrence on the other hand. Indeed, the Data Acquisition Layer allows the collection of information that is subsequently sent to the Data Processing and Storage Layer through the Connection system layer for processing and storage. The processed data are used to send alerts to different stakeholders through the Notification Layer by using services offered by the Service Layer.

In future work, our architecture will be improved to take into account all of the IoT ecosystem and new technologies, protocols and devices. We will provide full details on the implementation of each stack of our architecture.

References

1. Gubbi, J., Buyya, R., Marusic, S., Palaniswami, M.: Internet of Things (IoT): a vision, architectural elements, and future directions. Future Gener. Comput. Syst. **29**(7), 1645–1660 (2013)
2. Hanes, D., Salgueiro, G., Grossetete, P., Barton, R., Henry, J.: IoT Fundamentals: Networking Technologies, Protocols, and use Cases for the Internet of Things. Cisco Press, Indianapolis (2017)
3. FAO, Exemples de réussites de la FAO en matière d'agriculture intelligente face au climat (2014). www.fao.org/climatechange/29830-0beb869615795a9960ada6400c62b3783.pdf. Accessed Oct 2019
4. Kitchell, E., Turner, M.D., McPeak, J.G.: Mapping of pastoral corridors: practices and politics in eastern Senegal. Pastoralism **4**(1), 17 (2014)
5. AFDB, Bénin : La vallée de l'Ouémé se transforme grâce au PAIA-VO (2017). https://www.afdb.org/fr/projects-and-operations/selected-projects/benin-la-vallee-de-loueme-se-transforme-grace-au-paia-vo-23. Accessed Oct 2019
6. Marz, N., Warren, J.: Big Data: Principles and Best Practices of Scalable Realtime Data Systems. Manning, New York (2013)
7. Pawar, K., Attar, V.: A survey on data analytic platforms for Internet of Things, In: Proceedings of CAST, pp. 605–610 (2016)
8. Aloÿs, A., Jiazi, Y., Thomas, C., Townsley, W.: A study of LoRa: long range & low power networks for the Internet of Things. Sensors **16**(9), 1466 (2016)
9. Zanella, A., Bui, N., Castellani, A., Vangelista, L., Zorzi, M.: Internet of Things for smart cities. IEEE Internet Things J. **1**(1), 22–32 (2014)
10. Shelby, Z., Hartke, K., Bormann, C.: The constrained application protocol (CoAP). No. RFC 7252 (2014)
11. Locke, D.: MQ telemetry transport (MQTT) v3.1 protocol specification. IBM developerWorks Technical Library, p. 15 (2010)
12. Singh, M., Rajan, M.A., Shivraj, V.L., Balamuralidhar, P.: Secure MQTT for Internet of Things (IoT). In: Proceedings of Fifth CSNT, pp. 746–751 (2015)
13. Qiu, J., Wu, Q., Ding, G., Xu, Y., Feng, S.: A survey of machine learning for big data processing. EURASIP J. Adv. Signal Process. **2016**(1), 67 (2016)

14. Che, D., Safran, M., Peng, Z.: From big data to big data mining: challenges, issues, and opportunities. In: Hong, B., Meng, X., Chen, L., Winiwarter, W., Song, W. (eds.) DASFAA 2013. LNCS, vol. 7827, pp. 1–15. Springer, Heidelberg (2013). https://doi.org/10.1007/978-3-642-40270-8_1

15. Gandomi, A., Haider, M.: Beyond the hype: big data concepts, methods, and analytics. Int. J. Inf. Manage. **35**(2), 137–144 (2015)

16. Dittrich, J., Quiané-Ruiz, J.A.: Efficient big data processing in Hadoop MapReduce. In: Proceedings of VLDB Endowment **5**(12), 2014–2015 (2012)

17. Li, T., Liu, Y., Tian, Y., Shen, S., Mao, W.: A storage solution for massive IoT data based on NOSQL. In: Proceedings of IEEE GreenCom (2012)

18. Cai, H., Xu, B., Jiang, L., Vasilakos, A.V.: IoT-based big data storage systems in cloud computing: perspectives and challenges. IEEE Internet Things J. **4**(1), 75–87 (2017)

19. Arora, V., Nawab, F., Agrawal, D., El Abbadi, A.: Multi-representation based data processing architecture for IoT applications. In: Proceedings of 37th ICDCS, pp. 2234–2239 (2017)

20. Clanet, J.C., Ogilvie, A.: Farmer-herder conflicts and water governance in a semi-arid region of Africa. Water Int. **34**(1), 30–46 (2009)

21. Ange, M., Kinhou, B., Brice, S.: Transhumance and conflicts management on Agonlin plateau in Zou department (Benin). J. Biodivers. Environ. Sci. (JBES) **4**(5), 32–145 (2014)

22. Patil, K.A., Kale, N.R.: A model for smart agriculture using IoT. In: Proceedings of IEEE ICGTSPICC (2016)

23. Srinivasulu, P., Babu, M.S., Venkat, R., Rajesh, K.: Cloud service oriented architecture (CSoA) for agriculture through Internet of Things (IoT) and big data. In: Proceedings of ICEICE (2017)

24. Popović, T., Latinović, N., Pešić, A., Zečević, Ž., Krstajić, B., Djukanović, S.: Architecting an IoT-enabled platform for precision agriculture and ecological monitoring: a case study. Comput. Electron. Agric. **140**, 255–265 (2017)

Agrilogistics - A Genetic Programming Based Approach

Divya D. Kulkarni[✉] and Shivashankar B. Nair

Indian Institute of Technology Guwahati, Guwahati, India
{divyadk,sbnair}@iitg.ac.in

Abstract. The advent of technology in the agriculture sector, such as precision agriculture, the Internet of Things (IoT) and machine learning has dramatically improved the experience of farming scenario. Apart from improving the farming conditions, there is a need for focused effort to achieve a balanced ecosystem in the supply chain of agrilogistics. Inefficient price signals conveyed to the farmer, erratic price fluctuations and inflation of the agri-produce coupled with the presence of several intermediaries, tend to imbalance the system. In this work, we propose an IoT based agrilogistic system coupled with a genetic programming algorithm to ensure fair prices across all the participants within. The system evolves and generates a set of programs that, in turn, generates the selling rate for every participant in the supply chain in a manner that confers fairness.

Keywords: Internet of Things (IoT) · Genetic Programming · Agrilogistics · Supply chain

1 Introduction

The agriculture sector has seen enormous changes lately due to the advent of technology. An IoT plays a significant role in traditional precision agriculture, wherein the data is collected from a variety of connected devices and sensors, setup in the fields, thus improving both yield and productivity. IoT in agriculture mainly comprises setting up sensors across large tracts of fields, surveying and collating data mostly using either sensor networks or drones. Sensors used are varied in nature and include those that can sense temperature, humidity and soil-moisture [4]. Cameras have also been used as sensors to judge the quality of the plants [1]. Various solutions have been proposed to augment IoT with precision agriculture [3,9,11,14], but only a few have been actually implemented effectively [6,18]. Many have proposed the use of machine learning and data mining techniques for improving yield, controlling pests, managing soil, etc. [15,16] but actual implementations of only a few have shown decent results [5,7].

There are several other factors that need to be addressed in the making of a fair agriculture based ecosystem. For instance inefficient price signals, the

P. Pereira et al. (Eds.): SC4Life 2019, LNICST 318, pp. 83–96, 2020.
https://doi.org/10.1007/978-3-030-45293-3_7

presence of too many intermediaries and information asymmetries [2,12] can lead to a severe losses to a farmer.

In order to protect the producer from erratic price fluctuations and to handle the problem of inefficient price signals in agrarian countries like India, the government generally sets up a Minimum Support Price (MSP).

Though this may protect the farmers from incurring severe losses, it does not guarantee the best price for their produce mostly because they are not aware of the overall dynamics of the market. Presence of too many intermediaries such as middle-men, agents, and brokers between the point of production (farm) to the point of vending and consumption also causes erratic price fluctuations. The problem at times can cascade resulting in a gross wastage of the product due to high production and low demand. Governments also ensure to stock warehouses managed by them with the produce to control market trends. Such warehouses buy the produce from farmers at the MSP and endeavour to always maintain a buffer stock of the produce. It may happen that the quantum of produce stored exceeds the buffer stock. Under such conditions, the same is released to the market at a fair price. Such an increase or decrease in stock in the warehouse coupled with dynamically varying production and demands can create a large turbulence in profits made by the supply chain. Mau et al. [10] points out that the agriculture sector lacks efficient consumer response, which could guide the producer in the production of the goods based on the demands of the customer. It is reported in [17] that profits are mostly skewed away from the producers due to the presence of various intermediaries. In an agrilogistic system, ensuring a mechanism that profits all the participants is an uphill task. To handle this supply chain in agrilogistics (logistics system of the agriculture sector), this mechanism would need to be aware of all the connected dynamics of the market so as to advise each of the participants of the action to be taken by them.

One of the methods to solve the supply chain agrilogistics is to initially collect a huge amount data which naturally can be generated only over a large period of time, and then analyze it using various machine learning techniques including deep neural networks. This could mean huge losses (during the data collection interval) with hardly anything gained or learned. A mechanism wherein learning commences during the data collection phase itself would greatly ameliorate the problem. One of the ways this could be realized is by making programs evolve using Genetic Programming (GP) [8] based techniques during the course of data collection. Execution of such programs will help steer the system being controlled to a better level. In GP, the quality of a program is measured by a fitness score, which in an agrilogistic scenario could be the selling rate of the product at every stage. In addition, an IoT based agrilogistic supply chain system, data received from heterogeneous sensors could guide the evolution of such programs.

In this work, we propose a Genetic Programming (GP) [8] based strategy to learn to handle the dynamics of the supply chain and accordingly advise the participants on whom to sell their stock in a network of producers (farmers), warehouses and vendors so as to get the best price. The strategy takes into consideration the dynamically changing demand, the quantum of stock,

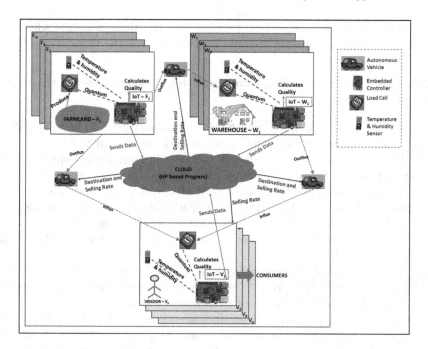

Fig. 1. The proposed model of supply chain management in agrilogistics

temperature, humidity and time elapsed after harvest and evolves programs over time which when run maximizes the selling rates across all the participants.

2   Methodology

The work herein describes a method to generate and evolve efficient programs for an IoT based supply chain of agrilogistics shown in Fig. 1. As illustrated the participants include sets of farmers, $F = \{F_1, F_2, F_3..., F_n\}$, warehouses, $W = \{W_1, W_2, W_3, ..., W_m\}$ and vendors, $V = \{V_1, V_2, V_3, ..., V_p\}$. A farmer F_i who generates the produce can either send them, part or whole, over to a warehouse W_j or a vendor V_k. Likewise W_j, which maintains a buffer stock of the produce, could channel excess amounts to any of the vendors based on demand. Eventually, the vendors sell the produce to the consumers. In addition, there are q number of number of autonomous vehicles (G) at a site x (where $x \in \{\{F_i\}, \{W_j\}, \{V_k\}\}$), G_{x_q}, at each of the locations or sites of the participants. A vehicle G is responsible for transporting a certain quantity of produce from one site to the other.

The IoT infrastructure is used to sense and record parameters, such as ambient temperature and humidity, age of the produce, its quantum and quality, at every site. The recorded data is then sent to the cloud which runs the proposed GP based algorithm to evolve programs, with the primary objective being to maximize the selling rates, while at the same time maintaining their consistency at the respective sites. The evolved program provides the destination where the produce has to be transported to and the selling rate at the destination.

Important Parameters. The main parameters and terms used in the system are explained below.

1. *Temperature (T) and Humidity (H):* Both temperature and humidity affect the quality of the produce. The measurements are done at the sites of every F_i, W_j and V_k using the temperature and humidity sensors at respective IoTs installed therein. Based on fixed predetermined ranges of values, temperature and humidity, are recorded as High (*hi*), Moderate (*mod*) and Low (*lo*).
2. *Quantum of Produce (Q):* It is the net quantity of a specific produce available at an instant of time at the site of a participant. Q also assumes values denoted as *hi*, *mod* and *lo*. It is measured using a load cell connected to the IoT at the respective sites.
3. *Age (τ):* The *age* of any produce is defined as the time elapsed from the time of harvest to the current time. The age of the produce increases irrespective of its location and is independent of all other parameters.
4. *Quality (ϕ):* The *quality* of the product affects its overall selling rate. Higher the *quality*, higher would be its selling rate. The quality of produce is assumed to degrade with increase in either T or H or τ. Just as temperature and humidity, ϕ of a produce takes values *hi*, *mod* and *lo*.
5. *Demand (δ):* It is one of the most significant terms and affects the dynamics of the system. δ increases when Q at the vendors is *lo* and decreases otherwise. Demand influences the production and outflux rates of the produce from a site, thereby affecting their selling rates at different sites that constitute the system. Demand also is expressed as *hi*, *mod* and *lo*.
6. *Distance (D) to the Destination (ψ):* It is the distance between the site from where the produce is to be shipped and the site where it is to be delivered (Destination site ψ).
7. *Outflux (O) of Produce:* Outflux is amount of produce a deliverer ships to the destination (ψ). It is calculated locally at the respective IoT site. In our work we assume the produce P to be *onions, tomatoes* and *cotton*.
   Outflux of the produce P from the site x to the site ψ is given by the formula below:
   $$O_P^{x \to \psi} = \alpha * \delta_P * Q_P^x \tag{1}$$
   where α is a constant and δ_P^x and Q_P^x are the demand and the quantum of the produce P at the site x, respectively and x and $\psi \in \{\{F\}, \{W\}, \{V\}\}$.
8. *Selling Rate (ρ):* The selling rate per unit of a produce, ρ, is determined at every site x for every produce P in the supply chain. The selling rate is directly proportional to the quality, ϕ, of the produce, demand from the customer, δ, and the distance, D, to the destination where it needs to be transported. It is inversely proportional to the quantum, Q, and age, τ, of the produce.

Selling rate, ρ, is given by the Eq. (2):

$$\rho_P^x = \beta * \frac{\delta_P * \phi_P^x * D_P^x}{Q_P^x * \tau_P^x} \tag{2}$$

where β is a constant, P is the produce whose ρ is being calculated at site x respectively, $x \in \{\{F\}, \{W\}, \{V\}\}$.

The parameters T, H and Q are sensed using the associated sensors while those of τ, ϕ, and O are computed by the respective IoTs, locally. The rest wiz. δ and ρ are computed in the cloud as shown in the Fig. 1.

Evolution of Programs. The generic version of a program in a population of programs, used in this work, has three tuples corresponding to a *farmer*, *warehouse* and *vendor* respectively, as shown below.

$$S \equiv ((F^{(T,H,P,\delta_P,\phi_P,Q_P,\psi,\rho_P)^i}), (W^{(T,H,P,\delta_P,\phi_P,Q_P,\psi,\rho_P)^j}), (V^{(T,H,P,\delta_P,\phi_P,Q_P,\rho_P)^k}))^l \tag{3}$$

where T, H, δ, Q, η can take either of the values *hi,mod* or *lo* and S is set of all such programs evolved. Each tuple of the program refers to all the *farmers*, all the *warehouses* and all the *vendors* at that instant of time. This tuple denotes the rule that infers the selling rate and the destination to which the produce needs to be shipped from the corresponding site i.e. F^i, W^j or V^k based on the values of the parameters. At the site, V^k, the produce is sold to the customers. For instance, the portion of the first tuple ($i = 1$) related to the first farmer's site of the program is illustrated below.

$$((F^{(lo,lo,onions,hi,hi,mod,W_2,45)})^1), (F^{(hi,lo,onions,hi,mod,mod,W_1,30)})^1)$$

The above portion of the program that can be expressed as a rule below -

FOR{*the site of the farmer F_1*} **IF** {*the temperature and humidity are low, the produce is onions with high demand having high quality and moderate quantum*}, **THEN** {*the selling rate is 45 units and the destination to which it needs to be shipped is W_2*} **ELSE IF** {*the temperature is high and humidity is low, the produce is onions with high demand having moderate quality and moderate quantum*} **THEN** {*the selling rate is 30 units and the destination to which it needs to be shipped is W_1*}.

Programs such as these are stored in a repository in a cloud. In the current scenario, the selling rate, ρ, of a produce is considered as the fitness score of the associated program. The goal is to maximize ρ so that the concerned seller get the maximum profit. It may be noted that the selling rate is contained by the demand of the consumers (as seen in Eq. 2) and hence programs evolve in a manner that is congenial to all participants, including the consumers.

The Algorithm

In the GP approach, the initial programs are generated randomly. In the work reported herein, it is proposed that the sensed parameters T, H, Q *and* ϕ are obtained and calculated using the IoT infrastructure at each of the sites participating in the agrilogistics scenario, thereby greatly curtailing the otherwise random search space. The GP program is assumed to be hosted in a cloud.

When all these values from all the sites are initially sent to the cloud, the demand of each type of produce is computed. Based on the demand, the outflux (O) of the produce is calculated. Using a randomly selected destination (ψ), the selling rate (ρ) of the produce (P) is then computed. These values viz. P, O, ρ and ψ are sent to the concerned site. These computations are done for every P at every site at the cloud. The respective IoT at the sites, communicates with an autonomous vehicle or a transporter within its site, which in turn delivers the O units of the produce P to the concerned destination, ψ. A program, as discussed earlier, which includes all the sites is thus generated and stored in a repository within the cloud. The cloud always receives the data from all the IoTs and structures the incoming data, I in the form shown below as a three tuple:

$$I \equiv (F^{\{T,H,P,\phi_P,Q_P\}^i}, F^{\{T,H,P,\phi_P,Q_P\}^j}, V^{\{T,H,P,\phi_P,Q_P\}^k})^l \tag{4}$$

The next time, the cloud receives and structures an I, the GP program first checks whether the first tuple for the first farmer, F_I^1, sent by the site matches with any one of the former portion of the first tuple F_S^1 in the set of stored programs S. If so, ψ_S^j (which could be either W_S^j or V_S^j) from the latter portion of F_S^1 is retrieved. Using this information, the GP program retrieves the tuple for the same in the second or third tuples, as the case maybe, in S. This retrieved tuple is compared with the corresponding tuple W_I^1 or V_I^1 (as the case maybe). The corresponding ψ_S^k (which could be either W_S^k or V_S^k) is retrieved. Using this information, the GP program retrieves the tuple for the same in the third tuples in S. This retrieved tuple is compared with the corresponding tuple W_I^1 or V_I^1 (as the case maybe). If all comparisons are true, then the three tuples from S namely F_S^1, W_S^j and V_S^k sent back to the IoT site of F_1. In case, no match is found, the GP evolves a new program using crossover and mutation as explained later.

The crossover operation is applied when the currently reported parameters match with at least one tuple in S stored in the repository. The operation of crossover involves two *parent programs* in S, the first *parent* is one whose former portion of the first tuple in S matches with the first tuple of I. The other parent in S could be one whose former portion of any of its two latter tuples matches with either of the second or third tuple of I. The crossover operation is illustrated in the Figs. 3, 4 and 5. If there are multiple candidates for the *parent programs*, then those *parents* are selected which infer the highest selling rates. The crossover operation results in two new programs are stored in the repository and retrieved as and when required in future. In this condition too, the three tuples from S namely F_S^1, W_S^j and V_S^k sent back to the IoT site of F_1.

If a few parameter values in F_I^1 do not match with the former portion of F_S^1, then those in F_I^1 are copied on to F_S^1 to achieve the mutated solution. Then, the corresponding selling rate is calculated and the solution is sent back to the IoT site of F_1. The mutation operation is illustrated in the Figs. 6 and 7.

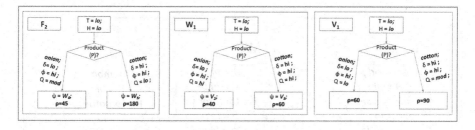

Fig. 2. Two of the programs generated at the initial generations corresponding to the items - **onion** and **cotton**

The process of matching, crossover and mutation is done across all tuples in I so as to complete one generation of evolution.

If there are no matching programs to perform either of the crossover or mutation operations, then a new program is generated and routed to a randomly selected destination with the selling rate determined as per the Eq. 2.

3 Results

Generation of Initial Population of Programs. We simulated the GP cum IoT side algorithms while generating the appropriate values of the pertinent sensors. First, we generated the initial population of programs as described in the previous section, and stored them in the repository within the cloud. Figure 2 shows the illustration of two of the initial the programs generated as per Eq. 3 for the products (*onion*) and (*cotton*). There are three blocks in the program viz. the farmer, the warehouse, and the vendor blocks representing the three tuples in Eq. 3, each corresponding to the conditions observed at the respective sites. In Fig. 2, the ρ corresponds to the fitness of that particular program at that site for the associated produce. The selling rate is calculated as per the Eq. 2. We generated 50 such programs as the initial population and stored them in the repository in the cloud.

Evolving New Programs. After the initial programs were generated over time, when the sites report the next set of observed conditions to the cloud, the GP program searches the repository for the matching programs as explained in the previous section. If the matching program is found, the cloud reports the ϕ where the O_P^x has to be routed, and ρ_P^x, at which the produce has to be sold to the ϕ. If a match is not found, then either the operation of crossover or mutation is applied.

The Figs. 3, 4 and 5 depict the scenario, where crossover operation is applied. Parent 1 program in Fig. 3 represents *Parent* 1, where the tuple corresponding to the site of the farmer F_2 match with the currently reported parameters of F_2, whereas the remaining tuples corresponding to W_1 and V_2, do not match with the currently reported parameter values. Similarly, *Parent* 2 program in

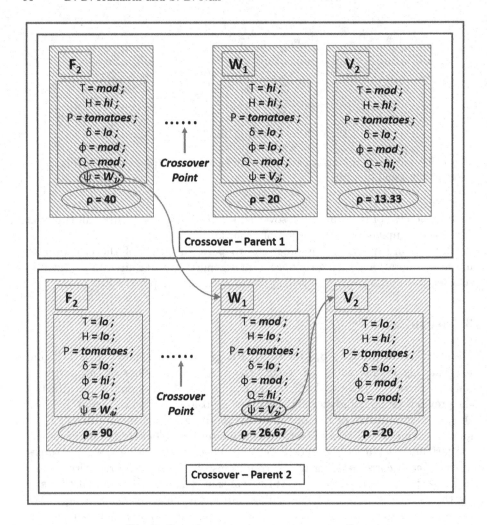

Fig. 3. Parent programs - crossover operation

the same Fig. 3 depicts *Parent 2* where the tuples corresponding to W_1 and V_2 match with the respective currently reported parameter values. The crossover mechanism is applied on the matched programs to obtain two new programs, one of them which can cater to the current parameters as shown in the Fig. 4 and the other one shown in Fig. 5 is saved in the repository for future use.

Mutation operation is depicted in Figs. 6 and 7. The *parent* program in Fig. 6, has the value of the parameter Q as *lo* for the site F_1, whereas the currently reported value of Q at F_1 is *mod*. This parameter is mutated as shown in the Fig. 7 and a new program is obtained. The selling rate (or the fitness) of this newly generated program is calculated and updated in the repository.

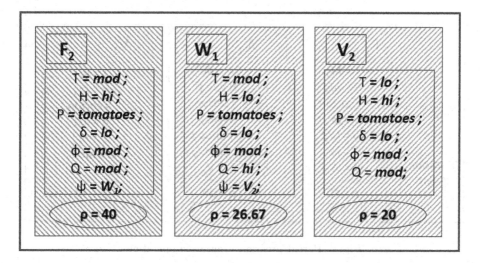

Fig. 4. Resultant program 1 - post crossover operation

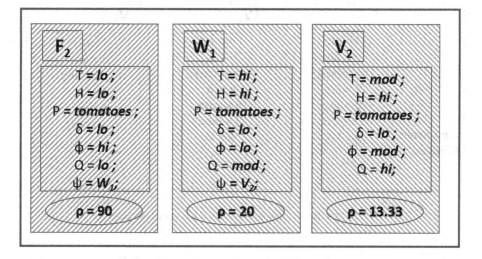

Fig. 5. Resultant program 2 - post crossover operation

As we generate and evolve the programs as per Eq. 3, the programs grow in dimension. The pseudocode in Pseudocode 1 and the Fig. 8, depicts the programs assembled for the site F_1 at an instant of time. As more and more programs are evolved, the conditions in the programs also increase, thereby resulting in a complex program. When we observe Figs. 2 and 8, we can observe the increasing conditions as more programs evolve.

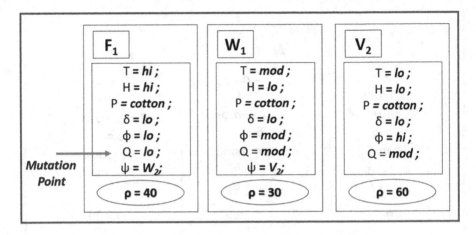

Fig. 6. Parent program - mutation operation

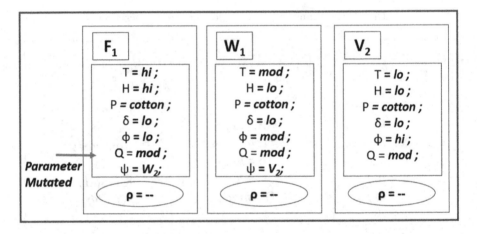

Fig. 7. Resultant program - post mutation operation

The graph of the selling rate of all the participants F, W and V plotted against the generations as the programs evolve is shown in the Fig. 9. The selling rates of all F, all W and all V is given by σ_F, σ_W and σ_V, respectively, as below:

$$\sigma_F = \sum_{i=1}^{m} \sum_{q=1}^{P} \rho_q^{F_i}$$

$$\sigma_W = \sum_{j=1}^{n} \sum_{q=1}^{P} \rho_q^{W_j} \tag{5}$$

$$\sigma_V = \sum_{k=1}^{o} \sum_{q=1}^{P} \rho_q^{V_k}$$

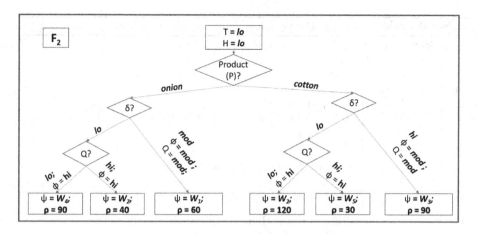

Fig. 8. A Snapshot of assembled programs

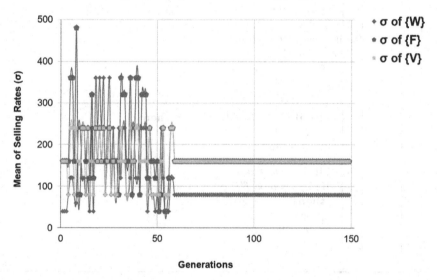

Fig. 9. Graph of the selling rates across generations

In the graph, shown in Fig. 9, the selling rates till 50th generation represent the initial population of programs generated. After which, the programs evolve as per the GP based algorithm. In the initial generations, we can observe fluctuations in the σ_F, σ_W and σ_V, wherein only one of the three sets of participants (F, W and V) seem to get high selling rates whereas others are getting lower values. However, as the programs evolve, all the σ converges to narrow bands. As can be seen, σ_F and σ_V converge to almost the same narrow band, while σ_W converges to a lower value. This is because, at each W_i there is certain amount of buffer stocked, causing an increase in the respective Q, consequently decreasing the ρ as per the Eq. 2. This lower value of σ_W conforms to the standard

Pseudocode 1 : Part of the pseudocode of one of the programs assembled

```
1  switch X do
2  │  case X ← F₂ do
3  │  │  if  T == lo && H == lo then
4  │  │  │  if P == onion then
5  │  │  │  │  if δ == lo then
6  │  │  │  │  │  if Q == lo && φ == hi then
7  │  │  │  │  │  │  ψ ← W₄
8  │  │  │  │  │  │  ρ ← 90
9  │  │  │  │  │  else if Q == hi && φ == hi then
10 │  │  │  │  │  │  ψ ← W₂
11 │  │  │  │  │  │  ρ ← 40
12 │  │  │  │  else if δ == mod then
13 │  │  │  │  │  if Q == mod && φ == mod then
14 │  │  │  │  │  │  ψ ← W₁
15 │  │  │  │  │  │  ρ ← 60
16 │  │  │  else if z == cotton then
17 │  │  │  │  if δ == lo then
18 │  │  │  │  │  if Q == lo && φ ← hi then
19 │  │  │  │  │  │  ψ ← W₂
20 │  │  │  │  │  │  ρ ← 120
21 │  │  │  │  │  else if Q == hi && φ == hi then
22 │  │  │  │  │  │  ψ ← W₅
23 │  │  │  │  │  │  ρ ← 30
24 │  │  │  │  else if δ == hi then
25 │  │  │  │  │  if Q == mod && φ == mod
26 │  │  │  │  │  then
27 │  │  │  │  │  │  ψ ← W₃
28 │  │  │  │  │  │  ρ ← 90
```

warehousing practices [13]. It can thus be seen that each participant in the supply chain gets a fair rate for the produce.

4   Discussions and Conclusions

We have proposed a novel way of looking at the Supply Chain in the agriculture sector, to setup fair prices of commodities by taking into consideration the factors which affect the selling rate. This will eliminate the influence of middlemen on the prices, and possibly bring down erratic price fluctuations. The proposed work combines an IoT with supply chain management in agrilogistics and gives an indication to the farmer as to what products are in demand enabling him/her to concentrate on producing them. This insulates the farmer from producing excessively large quantities of a low demand produce or vice versa which may otherwise result in wastage and financial losses. Such an end to end system, provides for a win-win situation for all the participants involved in the supply chain. This can be clearly seen from the resulting graph wherein the selling rate for all participants seem to converge to a narrow band. While the IoT provides a clear and exact picture of the ground truth of the produce at various sites, the GP learns from what it has made the system do in the past using the programs in its repository. In some sense it learns from scratch and does not really require for one to wait for the collection of large amounts of data before arriving at a solution. Most of the current AI methods bank on the initial availability of a huge amount of relevant data which is then churned to produce some meaningful results using computationally heavy algorithms such as deep learning. Learning from what has been done so far during the data collection phase is grossly missing. The paradigm discussed herein could thus be well suited for learning during a data collection phase. It may even allow for better and more pertinent data to be generated thereby eventually increasing the efficacy of the traditionally used large-data driven algorithms. There are yet certain other aspects which need to be addressed to optimize the transport of goods from one site to another as also put the IoT in the real-world. Our future work will thus include the realization of an actual agent based IoT and cloud to comprehend the challenges of the paradigm when implemented in the real-world.

References

1. Bock, C.H., Nutter Jr., F.W.: Detection and measurement of plant disease symptoms using visible-wavelength photography and image analysis. Plant Sci. Rev. **6**, 73 (2012)
2. Chand, R.: Development policies and agricultural markets. Econ. Polit. Wkly. **47**(52), 53–63 (2012). http://www.jstor.org/stable/41720551
3. Gebbers, R., Adamchuk, V.I.: Precision agriculture and food security. Science **327**(5967), 828–831 (2010). https://doi.org/10.1126/science.1183899. https://science.sciencemag.org/content/327/5967/828
4. Gondchawar, N., Kawitkar, R.: IoT based smart agriculture. Int. J. Adv. Res. Comput. Commun. Eng. **5**(6), 838–842 (2016)
5. He, Y., Zeng, H., Fan, Y., Ji, S., Wu, J.: Application of deep learning in integrated pest management: a real-time system for detection and diagnosis of oilseed rape pests. Mob. Inf. Syst. (2019). https://doi.org/10.1155/2019/4570808
6. Kapetanovic, Z., Vasisht, D., Won, J., Chandra, R., Kimball, M.: Experiences deploying an always-on farm network. GetMobile Mob. Comp. Commun. **21**(2), 16–21 (2017). https://doi.org/10.1145/3131214.3131220
7. King, A.: Technology: the future of agriculture. Nature **544** (2017). https://doi.org/10.1038/544S21a
8. Koza, J.R., Koza, J.R.: Genetic Programming: On the Programming of Computers by Means of Natural Selection, vol. 1. MIT Press, Cambridge (1992)
9. Mat, I., Kassim, M.R.M., Harun, A.N., Yusoff, I.M.: IoT in precision agriculture applications using wireless moisture sensor network. In: 2016 IEEE Conference on Open Systems (ICOS), pp. 24–29. IEEE (2016)

10. Mau, M.: Supply chain management in agriculture - including economics aspects like responsibility and transparency. In: European Association of Agricultural Economists, 2002 International Congress, 28–31 August 2002, Zaragoza, Spain (2002)

11. Mekala, M.S., Viswanathan, P.: A survey: smart agriculture IoT with cloud computing. In: 2017 International Conference on Microelectronic Devices, Circuits and Systems (ICMDCS), pp. 1–7, August 2017. https://doi.org/10.1109/ICMDCS.2017.8211551

12. Negi, S., Anand, N.: Supply chain of fruits & vegetables' agribusiness in uttarakhand (India): major issues and challenges. J. Supply Chain Manag. Syst. 4(1), 43–57 (2015)

13. Ozsen, L., Coullard, C.R., Daskin, M.S.: Capacitated warehouse location model with risk pooling. Naval Res. Logistics (NRL) **55**(4), 295–312 (2008)

14. Patil, K.A., Kale, N.R.: A model for smart agriculture using IoT. In: 2016 International Conference on Global Trends in Signal Processing, Information Computing and Communication (ICGTSPICC), pp. 543–545, December 2016. https://doi.org/10.1109/ICGTSPICC.2016.7955360

15. Patil, S.S., Thorat, S.A.: Early detection of grapes diseases using machine learning and IoT. In: 2016 Second International Conference on Cognitive Computing and Information Processing (CCIP), pp. 1–5. IEEE (2016)

16. Shekhar, Y., Dagur, E., Mishra, S., Sankaranarayanan, S.: Intelligent iot based automated irrigation system. Int. J. Appl. Eng. Res. **12**(18), 7306–7320 (2017)

17. TeamYS: Agri-logistics in India: challenges and emerging solutions (2013). https://yourstory.com/2013/05/agri-logistics-in-india-challenges-and-emerging-solutions/

18. Vasisht, D., et al.: FarmBeats: an IoT platform for data-driven agriculture. In: 14th USENIX Symposium on Networked Systems Design and Implementation (NSDI 17), pp. 515–529 (2017)

Real World Third-Person with Multiple Point-of-Views for Immersive Mixed Reality

Zebo Yang[⊠], Mingshu Zhang, Taili Zhang, Linhao Fu, and Tatuso Nakajima

Waseda University, Tokyo 169-8050, Japan
{zebo,momochi,ztl_eric,nogizakakazumi,
tatsuo}@dcl.cs.waseda.ac.jp

Abstract. Technologies of Virtual Reality (VR), Augmented Reality (AR) and Mixed Reality (MR) are increasingly applied in urban construction and smart cities. As one of the high fidelity revivification methods, the purpose of these applications is to provide a better immerse experience over the combination of realistic and virtual interactions. Numerous methods and models are adopted to simulate real life scenarios. Regularly, first-person perspective (1PP) is a spontaneous way to revive reality virtually, and this is also what most manufacturers do. In a different way, we tend to verify the feasibility of a new perspective for VR, AR or MR: a third-person perspective (3PP) with multiple Point of View (PoV) from different fixed cameras. We implemented an immersive AR environment for users to control or view themselves in a third person perspective. They are able to switch among different perspectives in different angles, including a first-person view. We developed a simple game based on this environment and conducted user study with this implementation. The result shows that people will generally be intrigued and willing to pay for a new MR experience (like 3PP) if the process is effortless and pleasurable. The most welcomed thing is the ability to switch perspectives during a VR or AR experience. This environment can be extended to real world services such as interviewing, dating and picnic, etc. or future smart cities services that use VR and AR, such as VR traffic management, VR communities empathetic planning and AR navigation, etc.

Keywords: Virtual Reality · Augmented Reality · Third person view · Future cities · Video games

1 Introduction

Today, the diversity of Virtual Reality (VR), Augmented Reality (AR) and Mixed Reality (MR) applications enters the field of urban settlements and smart cities [1, 2]. Prolificacy in similar patterns of AR and VR applications, coupled with constantly changed future possibilities produces a great demand of novel experiences. While AR and VR hardware are augmented from time to time, the design of Head-Mounted Display (HMD) spontaneously inspires the same pattern: first-person perspective (1PP) centered with the HMD. For example, Minecraft VR [3], an exploratory VR game with a low-poly aesthetic, Pokemon Go [4], an AR game for capturing Pokemon or battling gyms, Fragments

© ICST Institute for Computer Sciences, Social Informatics and Telecommunications Engineering 2020
Published by Springer Nature Switzerland AG 2020. All Rights Reserved
P. Pereira et al. (Eds.): SC4Life 2019, LNICST 318, pp. 97–108, 2020.
https://doi.org/10.1007/978-3-030-45293-3_8

[5], a mystery-solving MR game for Hololens, and researches like [6, 7], are all designed in the classic first-person view. However, first-person perspective sometimes cannot adequately tender horizon or information for virtual or augmented realities, especially in cases that require a macroscopic view, such as VR or AR tools for self-validation, rehabilitation, traffic detection, safety instruction or education [8–11]. We believe essential future smart cities services would be energized and consummated with these AR or VR enhancement.

With the question of perspective comparison [12] and the intention to explore potential AR/VR modes, we tend to probe into different applicable perspectives for MR users. Specifically, we verify the capability of fixed-camera third-person perspectives (3PP) for immersive AR applications and the ability to switch between different perspectives. In order to implement the fixed-camera third-person views for MR applications, we set up a rectangle room with cameras attached at different positions and use it as the experiment space. Each position of the cameras represents a different perspective for the user. The display of the VR HMD will show the live video footages from these cameras. When the user enters the room, he/she is able to see himself/herself from one of the perspectives created by these cameras. What the user does in the room represents what he/she operates to the virtual world and he/she is able to view all the actions from the HMD in real time. Moreover, we augment the video contents from the cameras, adding augmented and mixed realities to the environment. We set up action triggers in the room: when the user 'touches' one of these triggers, corresponding event will be operated. For example, when the user touches a locker, information of the locker owner will show up. Ultimately, content shown in the HMD will be automatically switched to another camera/perspective when the system detects the vision of the user may be blocked by his/her own body. It will be easier for the user to finish certain tasks in this way. In the case of Fig. 1, when the user steps through the trigger, the viewpoint will be switched to the right perspective from the left one.

This paper aims to create a third-person immersive environment for the upon features and verify the effectiveness of this perspective for AR, VR and MR. The prototype is built upon Unity 3D [13] and we use several webcams as the sources of different perspectives. Video stream of the right camera will be shown in HMD and serve as the current perspective. HTC VIVE [14] is used as the VR headset and TPCAST [15] is adopted to make the whole process wireless. TPCAST is a wireless adapter attached to the HMD, broadcasting the content from PC to the VR headset, removing the need for transmitting cables, which can also be done by the wireless adapter from VIVE [16].

Beyond entertainment, VR, AR and MR is likewise significant for many scenarios, especially for future smart cities services [10, 17]. Derivative applications of the third-person view environment can be of great help in these scenarios. For example, recent treatment approaches with virtual reality in stroke rehabilitation is rapidly adopted in clinical settings [18], and virtual reality researches on urban safety for preventing child pedestrian injury come into notice [10]. With the perspectives of multiple Point of View (PoV), in future cities, users may be able to observe themselves in a third-person view timely, anywhere and anytime, if the involved devices become perfectly wireless and portable, or the city gives the effort to open the CCTV (Closed-Circuit Television) for limitary public access.

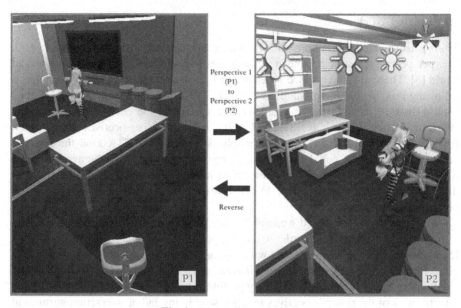

Fig. 1. Perspective switching

In the following sections, Sect. 2 describes several related works. Section 3 introduces the prototype implementation. Section 4 proposes the design and results of the user study. Section 4 describes limitations and challenges. Lastly in Sect. 5, we give conclusion and future possibilities.

2 Background and Motivation

In terms of the great attention paid to the ability of VR to immerse people into another reality, it seems to imply that first-person perspective is the perfect option for VR. As more and more AR or MR headsets come into mass production, such as HoloLens 2 [19], the ability to augment reality is also attached to the first-person perspective. We believe it can be more than like that. Playing in multiple perspectives in mixed realities is not always a second rate experience. For example, Chronos [20] uses a third-person perspective in the VR game by following the protagonist in a third-person camera. In many cases, it is more comfortable for a user to keep distance from the action. In this way, the user may feel less of the VR Sickness [21]. Based on the products of first-person and third-person view with a virtual main character, we tend to research and verify the combination of multiple perspectives in AR/VR. Further, we make the virtual protagonist real by using real-time video feeds from a fixed-camera, creating an immersive mixed reality in a real world third-person perspective: the protagonist is you and you are in a third-person point of view. To verify this two aspects, we developed a prototype and conducted user study with it.

2.1 Related Work

Third-person perspective has appeared in video games and AR/VR applications for years. This perspective seems preferable in action games with a protagonist roaming in a virtual world [3]. There are roughly two types of third-person perspectives in video games: a following PoV behind the protagonist like GTA [27], or a fixed-camera PoV attached from a distance, containing the character and a range of the environment where the character operates in, such as Biohazard 2 [28]. All of these perspectives are important for video games, and in [4], the authors check whether the GTA kind third-person perspective in AR/VR could share a preference. We tend to verify the availability of fixed-camera PoV with a perspective switching strategy: third-person perspectives from different angles (as shown in Fig. 1) and a first-person perspective (the classic AR perspective like Hololens 2).

A research of vision-based wearable device proposes dynamic gestures to interact virtual objects in the scene, in order to create a pervasive and interactive AR experience [5]. It provides a new way to interact with objects in augmented reality, using technologies of image processing to recognize different gestures. It inspired us to adopt similar solutions to indoor localization and body gesture detection. Since we tend to verify the effectiveness of the new perspectives in this research, instead of interaction gestures at this moment, we did not implement gesture based interaction in our prototype. Instead, we use VIVE Tracker [22] to implement an approximate indoor localization and use Vuforia Engine in Unity [23] to recognize markers in the scene, which play the role of the perspective switching and action triggers.

In [8], authors mention the technological impact of third-person perspective on daily life and its adoption on therapy and rehabilitation, especially coupled with virtual reality. We share the same idea on the impact of VR, AR and MR on the case of medication. In our future works, we plan to adapt our implementation to the scenario of medication and rehabilitation and verify the effectiveness of fixed-camera third-person view in this way. The implementation of this research is compatible with most indoor situations with proprio-perception intention, such as self evaluation of dancing, exercising, doing work out or other behaviors. We believe multiple perspectives for VR and AR have incremental weight on diversifying people's daily lives in future smart cities.

2.2 Motivation

As more AR/VR applications and more gameplay modalities emerge due to the demanding market, people may suffer from the gap between themselves and the points of view in the virtual or augmented reality. For example, a fast action will be quite difficult to track on a standard first-person perspective or a classic following third-person view [4]. Manufacturers will probably need some extra effort before adding this category of virtual ingredients to the virtual or augment reality. On the other hand, when inventing a teleport characteristic for the protagonist, such as teleport between dimensions in Halo 5 [24], the suddenly switch of camera to a different context is easy to leave a sense of disorientation. A fixed camera third-person perspective does not have these problems. Within the current field of view, it is spontaneous to implement any fast action or teleport features. With a fast pace and fiercely competitive AR/VR market, one more effective choice is always safe to have. Thus, we want to verify the availability of this perspective and multiple PoV in VR, AR and MR implementations.

To tackle this, we create a mixed reality environment with multiple selectable PoV, including first-person perspective and fixed camera perspective from different angles. These points of view are implemented as camera footages streaming to a wireless VR HMD. With this environment, we tend to verify two things: the acceptance of fixed camera third-person perspective in virtual reality and the preference of the ability to change perspective in virtual reality.

3 Design and Implementation

In order to verify the availability of fixed camera third-person perspective and the ability of selectable PoV, we implemented a prototype based on Unity 3D and HTC VIVE, and invited people to try it and participate in the experiment. In this section, we introduce the settings of hardware and software configuration of our prototype.

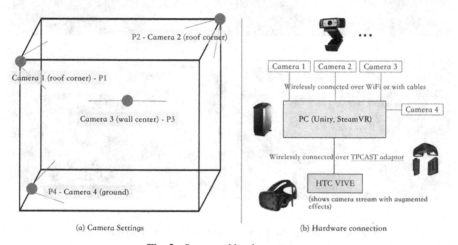

(a) Camera Settings   (b) Hardware connection

Fig. 2. Space and hardware setup

3.1 Hardware Setup

We set up a rectangle room (As shown in Fig. 2a) as the experiment space, which is covered by the field of view of four cameras. Two of the four cameras are attached diagonally to the corner of the roof while one of the rest is attached to the middle of the longer wall and the last one is set on the ground of the shorter side. With these four cameras, we simulate four different PoV from different angles. As shown in Fig. 2b, all of these cameras are connected to a PC that runs the Unity project. Footage of the current camera selected by the user will be processed by Unity and eventually streamed to the HTC VIVE headset. We use Logitech Wide Angle Webcam and Broadcaster Wi-Fi Webcam [25] for these perspectives.

The selection of these four camera positions corresponds to the four angles that video games frequently used in fixed camera third-person perspective. The correlation between these four positions and perspectives is shown on Fig. 3. The two cameras at the corners of the roof are considered to have the widest range of FoV (Field of View) and contain the most information in the space. The one at the middle of the wall provides a horizontal view and the one at the ground looks up at the user and provides the information under any covers like the table top or upper side of the shelf, which is blocked when using the looking down cameras.

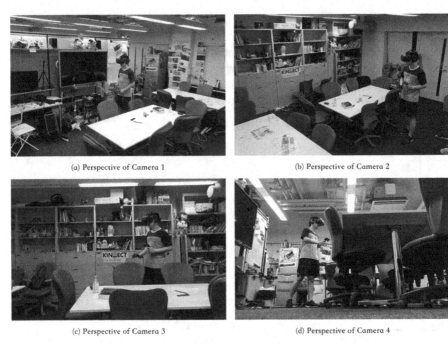

(a) Perspective of Camera 1

(b) Perspective of Camera 2

(c) Perspective of Camera 3

(d) Perspective of Camera 4

Fig. 3. Correlation between the four positions and perspectives

On the other hand, considering the flexibility of the experiment process, we need to allow the user to move freely in the room. This require the headset to be wirelessly connected, instead of letting it drag behind or probably trip over the user. In order to effectuate a wireless headset, we originally tried to process the Unity game or the PC desktop as video stream and project it to a smart phone and use Cardboard [26] to visualize the VR content. However, this process is laborious and turned out with bad performance because of the limited computing power of a smart phone. Instead, we use a third-party gadget called TPCAST to wirelessly transmit VR content to HTC VIVE headset with an integrated router, a transmitter and a receiver, substituting the cable transmission. Likewise, another option could be the wireless adapter from VIVE [16].

We set up markers in the room for mixed reality effects. The markers serve as the interactive objects in the game and are rendered as virtual objects like treasures, keys or any other necessary props in the virtual world. Users can interact with these objects with the VIVE controllers or directly touch them. Besides, there are some virtual triggers and objects in the game without physical markers for implementing virtual interactive objects, for example, a moving ghost. In future works, we may also use markers to set up some critical triggers in the room. Perspective will be automatically switched when one of these triggers is touched off. The positions of triggers and the way to touch off them should be selected and tested discreetly and deliberately because a wrong trigger would be a huge shock for the user.

3.2 Implementation

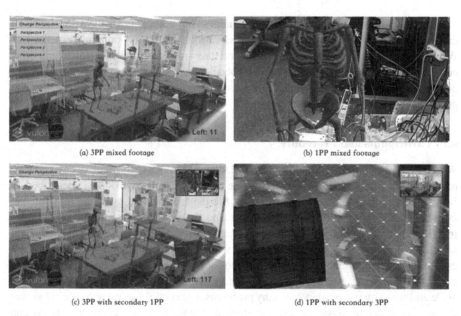

(a) 3PP mixed footage

(b) 1PP mixed footage

(c) 3PP with secondary 1PP

(d) 1PP with secondary 3PP

Fig. 4. The game scene mixed with camera footage

Based on the above hardware configuration, we developed a simple game with Unity 3D and run it with HTC VIVE. The game is designed simple for proof of concept. The user is supposed to avoid a moving skeleton and get to the treasure box in a limited time. As shown in Fig. 4, we reconstruct a rectangle space of our lab into a Unity scene and make some of the objects in the space appear with virtual effects. The skeleton is a trigger in game moving back and forward in front of the start point and the treasure box is a marker in game placed behind the skeleton. If the user collides with this skeleton or does not finish the game within the limited time (120 s), the game is over and shows a failure interface with a restart button. If the user goes through the skeleton and touches the treasure box with the VIVE controller, the game is completed successfully and a success interface shows up. The virtual effects to the reconstruction of the experiment space is raw and unripe, but it proofs the feasibility of virtual reconstruction and the availability of the augmented real life experience. To guide the user through the experiment, we let one of our researchers stand inside the FoV of the cameras. He shows up in the game scene and serve as a RPG (role-playing games) instructor.

By default, we provide the camera 1 (fixed camera third person perspective (3PP)) perspective in the experiment. To finish the task, the user can switch to first-person perspective (1PP) or other angles of the third-person perspectives manually. We encourage users to use all of these perspectives, and compare the feedback of different angles and views. Users can change the current perspective by clicking on the buttons listed at the left top side of the screen (Fig. 4a) with the VIVE controller.

4   Initial User Study

In order to conduct an evaluation and assessment to our approach, we have developed an experiment that measures user scores based on various indicators of our implementation. We asked participants to do questionnaires and interviews after the experiment and drew conclusion from the statistical result and feedback records. We aim to verify the approach in two respects: the fixed-camera third-person view for MR and the ability to switch perspectives in an immersive experience.

4.1   Participants and Procedure

We invited 12 college students (age m = 23.80, SD = 1.84, six men, six women) to participate in the user study. We set up the demonstration environment described in Sect. 3, and helped the participant to reset the experiment and wear the headset. We introduced the experiment, the perspectives and the objective task to the participants individually, and taught them how to interact with virtual objects. We showed them how to switch perspective and how to deal with triggers and markers. Participants are encouraged to use as many perspectives as possible before finishing the experiment and asked to successfully finish the task for at least one time. We observed the participants aside and stood by in case they have any problems. Users could actually see one observer in the game scene because the observer stood inside the FoV of the cameras. Users can turn to this 'RPG character' for instructions, just like interacting with virtual characters in classic role playing games, for example, GTA 5 [27]. Each participant was asked to fill out a questionnaire after the experiment, followed by a one-on-one interview.

4.2   Results and Feedback

We adopted the five-point Likert scale (1 - disagree/bad, 5 - agree/good) to develop ten rating queries. We asked the participants to complete the questionnaire right after the experiment. We observed that each experiment cost about 10 min averagely and the questioning and interview cost roughly 15 min for each participant. From the result of the questionnaire (as shown in Table 1), we know 91.7% of the participants have used VR devices before. The participants rated highest for the ability of changing perspectives in an immersive process (83.3% rated over 3 points).

From Table 1, we know P1 and P2 cameras (diagonal roof corners in Fig. 2) are the most welcomed perspectives among the four fixed camera third-person views in the

Table 1. Result of rating questions. (P1–P4 cameras are explained in Fig. 2)

	Average points	Standard deviation
Classic 1PP VR	3.83	0.58
Classic 1PP AR	3.92	0.67
Fixed camera 3PP (P1 + P2) in the demo	3.68	0.99
Fixed camera 3PP (P3) in the demo	3.61	0.90
Fixed camera 3PP (P4) in the demo	3.42	0.89
Fixed camera 3PP (Overall) in the demo	3.57	0.90
1PP AR in the demo	4.42	0.51
Able to switch perspectives	4.50	0.74
3PP with secondary 1PP (Fig. 4c)	3.87	0.63
1PP with secondary 3PP (Fig. 4d)	4.49	0.79
Generally about the experience	4.00	1.02
Willing to pay if well-developed	3.75	0.62

prototype, but collectively lower than the traditional first-person perspectives. However, when the fixed camera 3PP is combined with the classic 1PP, the ratings generally become higher. Specifically, rating of 1PP with secondary 3PP (Fig. 4d) is significantly increased and even surpass the classic VR and AR 1PP.

Moreover, P3 camera is rated slightly lower than P1 and P2 while P4 is the lowest. Overall, fixed camera third-person perspective is no better than the classic first-person perspective in AR/VR, but the P1 and P2 perspectives are generally accepted and could be a practical extra perspective for VR or AR applications. For example, use fixed-camera third-person views in indoor occasions and first-person in outdoor environments, having the secondary first-person and third-person attached correspondingly. Also, the rating for the ability of changing perspective during a gameplay is high. We speculate that people enjoy the freeness of selecting their own preferable perspective, instead of having what a system defines for them.

On the other hand, from the answers of multiple selection questions and interviews with the participants, we find that most participants (83.3%) consider the ability to switch perspectives is the best feature of this immersive MR environment and as a fresh experience, this third-person view is generally novel and acceptable to them. 66.6% of them are willing to pay for applications with such perspectives involved if the applications are fully developed. The most critical problem of this environment is the unnatural perspective. Users are worried that they may feel dizzy in this perspective if playing with it for long. 'Easy to block the vision' and 'hard to control' are ones of other concerns. We select the most typical ones and list them in Table 2.

Based on the result, we speculate that immersive third-person has incremental values to VR or AR applications, especially when it is combined with a natural perspective and serve as a supplementary perspective, providing more information and feedbacks for the users. The participants generally approve of the concept of this MR experience

Table 2. Pros and Cons.

Type	Description	Percentage
Pros	Be able to switch perspectives	83.3%
Pros	A fresh experience with AR/VR	41.7%
Pros	Be able to observe yourself	62.3%
Cons	May feel dizzy because of the unnatural perspective	58.3%
Cons	Easy to block the vision behind you	33.3%
Cons	Hard to control	33.3%

(4.00 points out of five in Table 1) and we believe the acceptability of such products is dependent on the experience, operability, how entertained the game is or how effective the application is, which demands reliable and fail-proof future implementations.

5 Current Limitations and Challenges

From the fixed camera 3PP demonstration, we see possibilities and limitations of this perspective and we summarize the challenges of similar kinds of implementation. The most critical problem of the fixed camera 3PP is the unnatural view of oneself. Users will hesitate to move at the beginning because they feel unreliable in an unnatural perspective. They are afraid to collide something unintentionally and feel hard to control themselves. A possible scenario is to use 1PP mainly and give a smaller rectangle view of the fixed camera 3PP at the corner as a secondary perspective (Fig. 4d). Users can use this extra perspective to obtain more information and don't have to stay in an unnatural perspective. They can switch the perspectives if they get used to the fixed camera 3PP and want to have a more macroscopic view.

On the other hand, fixed camera 3PP only provides one side view of objects or people in the room, so if you can see the front, you can not see the back, and vice versa. To overcome this constrain, users will have to switch between the diagonal perspectives to see the whole thing. The actions of switching between perspective can be automated and triggered at proper conditions such as when the user turn around, when the user's face is blocked by a table surface (when the user is trying to look below the table), or track the user's eyesight. This automation can be a challenge on development involved with image processing, face/motion detection, real time video processing and machine learning.

Compared with the following 3PP, fixed camera 3PP is more suitable for a complementary macroscopic view, instead of a sole perspective. Following 3PP is more spontaneous while multiple fixed camera 3PPs contain more information. Users will always prefer more customized options. Multiple fixed camera 3PPs will give incremental value to immersive experience when united with other perspectives, but they have difficulty serving as the only perspectives. Generally, people will get uncomfortable if using these perspectives for a long time. Solutions to avoid this uncomfortableness could be a challenge. One possible method is to combine these perspectives to a natural perspective like 1PP or following 3PP.

Moreover, this approach adopts commonly-used webcams to explore the efficacy of gaming perspectives for real world scenarios while leaving the camera variables out. The influence of camera variables in real life scenarios such as concave or convex viewpoints, mirroring effects and coverage issues, etc. could be explored accordingly.

6   Conclusion and Future Direction

We present a mixed reality indoor environment with multiple fixed camera third-person perspectives and a first person perspective for AR/VR applications, which can be extended to outdoor environments in future smart cities with ambient accessible intelligent cameras. We implement a prototype and conduct user study with this environment. Through the process, we speculate that immersive third-person has incremental values to VR, AR or MR applications. We consider the feasibility of such applications is up to their maneuverability, entertainment and effectiveness, which requires more fail-proof future researches. Basically, people will generally accept and be willing to pay for the new VR experience (fixed-camera third-person) when the process is effortless and pleasurable. We also conclude limitations and challenges of our implementation. A critical problem of this perspective is the unaccommodated view, which proof this perspective is not suitable for long-term use. Nonetheless, these perspectives provide a more macroscopic information of the environment and fresh angles for observing oneself. It is beneficial to combine fixed camera third-person view with classic first-person or following third-person perspectives. Additionally, the ability to choose a favorite perspective is generally a good option. In a such background of highly diverse immersive experiences, a novel perspective or angle will invariably be a fair alternative.

In the future, the interaction of perspective switch would be refined by motion detection such as detecting motions of raising head or pointing to the camera that provides the perspective you want to use. The game plot should be enriched and real people could show up and interact with users in the room, acting as interactive non-player characters (NPC). Beyond entertainment, we would verify the possibilities of multiple fixed camera 3PP usage in rehabilitation, education, self behavior analysis like dressing, dancing, etc., or smart cities services such as traffic management, urban safety instruction, etc. We want to adapt the 3PP environment to outdoor scenario with newly set-up or existing cameras (e.g. CCTV) and check on the potential derivatives.

References

1. Boris, P., Srdan, K., Maja, P.: Augmented reality based smart city services using secure IoT infrastructure. In: 28th International Conference on Advanced Information Networking and Applications Workshops, Victoria, BC, pp. 803–808 (2014). https://doi.org/10.1109/waina.2014.127
2. Zhihan, L., Tengfei, Y., Xiaolei, Z., Houbing, S., Ge, C.: Virtual reality smart city based on WebVRGIS. IEEE Internet Things J. 3(6), 1015–1024 (2016). https://doi.org/10.1109/JIOT.2016.2546307

3. Richard III, R.: What's your perspective? SIGGRAPH Comput. Graph. **33**(3), 9–12 (1999)
4. Patrick, S., Daniel, T., Frédéric, V.: The benefits of third-person perspective in virtual and augmented reality? In: VRST 2006 (2006)
5. Zhihan, L., Alaa, H., Shengzhong, F., Shafiq ur, R., Haibo, L.: Touch-less interactive augmented reality game on vision-based wearable device. Pers. Ubiquit. Comput. **19**, 551 (2015)
6. Slater, M., Spanlang, B., Sanchez-Vives, M.V., Blanke, O.: First person experience of body transfer in virtual reality. PLoS ONE **5**(5), e10564 (2010). https://doi.org/10.1371/journal.pone.0010564
7. Bergström, I., Kilteni, K., Slater, M.: First-person perspective virtual body posture influences stress: a virtual reality body ownership study. PLoS ONE **11**(2), e0148060 (2016). https://doi.org/10.1371/journal.pone.0148060
8. Patrick, S., Tej, T., Olaf, B., Frederic, V., Daniel, T.: Quantifying effects of exposure to the third and first-person perspectives in virtual-reality-based training. IEEE Trans. Learn. Technol. **3**(3), 272–276 (2010)
9. Barsom, E.Z., Graafland, M., Schijven, M.P.: Systematic review on the effectiveness of augmented reality applications in medical training. Surg. Endosc. **30**(10), 4174–4183 (2016). https://doi.org/10.1007/s00464-016-4800-6
10. David, C.S., Joanna, G., Joan, S.: Validation of virtual reality as a tool to understand and prevent child pedestrian injury. Accid. Anal. Prev. **40**(4), 1394–1400 (2008)
11. Jeremy, B., Kayur, P., Alexia, N., Ruzena, B., Sang-Hack, J., Gregorij, K.: The effect of interactivity on learning physical actions in virtual reality. Media Psychology **11**(3), 354–376 (2008)
12. Denisova, A., Cairns, P.: First person vs. third person perspective in digital games: do player preferences affect immersion? (2015). https://doi.org/10.1145/2702123.2702256
13. Unity 3D. https://unity.com. Accessed 20 June 2019
14. HTC VIVE. https://vive.com. Accessed 20 June 2019
15. TPCAST Wireless Adaptor. https://www.tpcastvr.com. Accessed 20 June 2019
16. HTC VIVE Wireless Adapter - Virtual Reality. https://www.vive.com/eu/wireless-adapter/. Accessed 20 Juen 2019
17. Lotfi, A., Faten Ben, A., Aref, M.: In-vehicle augmented reality traffic information system: a new type of communication between driver and vehicle. Procedia Comput. Sci. **73**(2015), 242–249 (2015)
18. Laver, K.E., Lange, B., George, S., Deutsch, J.E., Saposnik, G., Crotty, M.: Virtual reality for stroke rehabilitation. Cochrane Database Syst. Rev. **11**, CD008349 (2017)
19. Microsoft Hololens 2. https://www.microsoft.com/en-us/hololens. Accessed 20 June 2019
20. Gunfire Games, Chronos. https://www.oculus.com/experiences/rift/929508627125435/. Accessed 20 June 2019
21. Lee, J., Kim, M., Kim, J.: A study on immersion and VR sickness in walking interaction for immersive virtual reality applications. Symmetry **9**, 78 (2017)
22. HTC VIVE Tracker. https://www.vive.com/eu/vive-tracker/. Accessed 20 June 2019
23. Vuforia Engine for Augmented Reality. https://developer.vuforia.com/. Accessed 20 June 2019
24. Microsoft Studios, Halo 5: Guardians. https://www.microsoft.com/en-us/p/halo-5-guardians/brrc2bp0g9p0. Accessed 20 June 2019
25. Logitech Webcam. https://www.logitech.com/en-us/video/webcams. Accessed 20 June 2019
26. Google Cardboard - Virtual Reality. https://vr.google.com/intl/en_us/cardboard/. Accessed 20 June 2019
27. Rockstar Games, Grand Theft Auto V. https://www.rockstargames.com/V/. Accessed 20 June 2019
28. Capcom, Resident Evil 2. https://www.imdb.com/title/tt0161941. Accessed 20 June 2019

Internet of Things for Enhanced Smart Cities: A Review, Roadmap and Case Study on Air Quality Sensing

Gonçalo Marques[1,2(✉)] [iD] and Rui Pitarma[1]

[1] Polytechnic Institute of Guarda, Guarda, Portugal
goncalosantosmarques@gmail.com, rpitarma@ipg.pt
[2] Instituto de Telecomunicações, Universidade da Beira Interior, Covilhã, Portugal

Abstract. The smart city concept is a strategy to face relevant open issues regarding the socio-economic development requirements that promote people health and well-being. Smart homes are a fundamental element of smart cities and are intended to meet several applications for enhanced living environments and occupational health. Most individuals stay most of their time indoors, and therefore, smart homes are unquestionably an essential place to monitor. Indoor air quality has a meaningful negative influence on occupational health and well-being. Moreover, cities are responsible for a relevant portion of greenhouse emissions, and the outdoor air quality monitoring is relevant to detect adverse air quality situations and design interventions to promote public health. Air quality sensing is a fundamental element of people's daily routine and must be incorporated in smart homes and smart cities to promote health and well-being. Furthermore, this data can be evaluated by health professionals for diagnostics support and to correlate patient symptoms with their living environment but also to support city managers in the intervention planning for enhanced living environments. This paper presents a review and roadmap on air quality sensing and proposes an Internet of Things architecture for air quality supervision. The outcomes are encouraging as the presented system can be used to provide a correct and cost-effective air quality assessment. The proposed method can be used to support city managers to detect possible unhealthy scenarios in smart cities in useful time.

Keywords: Air quality sensing · Ambient assisted living · Enhanced living environments · Indoor environment quality · Internet of Things · Occupational health · Smart cities

1 Introduction

The Internet of Things (IoT) is a ubiquitous behaviour of a variety of entities with intelligence and sensing abilities which can collaborate to achieve shared objectives [1, 2]. IoT technologies offer numerous advantages to several domains such as smart cities, smart homes, healthcare and ambient assisted living (AAL) [3, 4].

© ICST Institute for Computer Sciences, Social Informatics and Telecommunications Engineering 2020
Published by Springer Nature Switzerland AG 2020. All Rights Reserved
P. Pereira et al. (Eds.): SC4Life 2019, LNICST 318, pp. 109–121, 2020.
https://doi.org/10.1007/978-3-030-45293-3_9

The smart city can be assumed as an approach to embrace modern urban productivity requirements in a general framework and to focus on the role of information and communication technologies (ICTs) on the creation of enhanced living environments [5]. Cities face relevant open issues regarding the socio-economic development requirements to promote people health and well-being [6]. Moreover, the smart city aims to moderate the obstacles caused by the urban population increase and accelerated urbanisation [7]. One of the most consistent open issues regarding smart cities is the no operability between different systems. IoT can present the interoperability to develop a centralised urban-scale ICT framework for enhanced living environments [8]. Is relevant to mention that smart cities will cause several meaningful effects at distinct layers on science, productivity and technology. The smart city will lead to complex challenges in society regarding the ethical and moral context. The smart city must provide correct information access to the right individuals. The security and privacy of that information must be established because that data is accessible at a distinct spatial scale where people can be distinguished [9].

Furthermore, IoT brings numerous opportunities regarding the development of modern daily routine applications and services for smart cities [10]. Several technologies closely related to the smart city context connected with IoT architecture will improve our daily routine and promote health and well-being [11].

Most individuals stay most of their time indoors, and therefore, smart homes are unquestionably an essential domain regarding the smart city context. IoT architectures can provide ubiquitous and pervasive methods for environmental data acquisition and provides wireless communication technologies for enhanced connectivity. Particularly, indoor air quality (IAQ) has a material adverse impact on occupational health and well-being. Therefore, the IAQ supervision is determinant for enhanced living environments and should be a requirement of all buildings and consequently, an integral part of the smart city context. The Environmental Protection Agency (EPA) recognises that IAQ values can be one hundred times higher than outdoor pollutant levels and established air quality on the top five environmental hazards to well-being [12].

Moreover, IAQ affects the most underprivileged people worldwide that remain exposed and therefore, can be compared with other public health problems such as sexually transmitted infections [13]. IAQ supervision must be seen as an effective method for the study and assessment of occupational health and well-being. The IAQ data can be used to detect patterns on the indoor quality and to design interventions plans to promote health. The proliferation of IoT devices and systems makes it possible to create automatic machines with sensing, connection and processing capabilities for enhanced living environments.

Air quality sensing must be an essential element of smart cities and smart homes. On the one hand, the IAQ can be estimated by providing CO_2 supervision system for enhanced public health and well-being. The CO_2 indoor levels can not only be used to evaluate the necessity of fresh air in a room but also can be used as a secondary sign of unusual concentrations of other indoor air pollutants. On the other hand, cities are responsible for a relevant portion of greenhouse emissions. Therefore, CO_2 monitoring must be provided on both indoors and outdoors environments. The concentrations of CO_2 are growing to 400 ppm, achieving further records every time after they started to be analyzed in 1984 [14]. The outdoor CO_2 monitoring can be relevant to plan interventions on traffic and to detect the emission of abnormal amounts of CO_2 in real-time. Moreover, real-time monitoring of CO_2 levels in smart cities at different locations allows the identification of points of interest to plan interventions to decrease the greenhouse gases.

This paper presents not only a review and a roadmap on the development of new methods and architectures for the implementation of air quality monitoring systems but also a case study regarding the development of a CO_2 monitoring system based in IoT architecture named **MoniCO₂** for enhanced living environments and public health in smart cities. This solution provides a mobile application for CO_2 data analytics and visualisation.

The rest of the document is written as follows: Sect. 2 presents a review on air quality monitoring systems; Sect. 3 is concerned to the application of air quality sensing in the smart city context; Sect. 4 presents the design of a cost-effective monitoring system for air quality assessment; Sect. 5 describes the discussion, and the conclusion is presented in Sect. 6.

2   A Review on Air Quality Monitoring Systems

Several IAQ management systems are presented in state of the art, a review summary is presented in this section. Numerous low-cost IoT methods for air quality management that supports open-source and wireless communication for data collection and transmission but also allow different places supervision at the same time through mobile computing technologies are proposed by [15–22].

An IAQ supervision system based on IoT, which incorporates temperature, humidity and CO2 sensing features developed for the smart city context is proposed by [23]. The data is collected and transmitted from the receiver node and is recorded on a personal computer. The data can be accessed through a LabVIEW and Android application.

An Air quality observation system based on the Arduino UNO microcontroller and low-cost sensors for CO_2 monitoring was proposed by [15]. The nodes receive and send data using ZigBee communication technology. However, this solution does not incorporate mobile computing technologies for data consulting.

The development of a wireless sensor network (WSN) for IAQ management is presented in [24]. This method incorporates temperature, humidity, gas and particulate matter (PM) sensors to determine the environmental health of monitored space. The collected data is used as feedback to control heating, ventilation and air conditioning systems in a smart building.

An IoT method to implement an IAQ management is presented in [25]. The proposed architecture incorporates portable sensors for data collection and implements a low power area network. The collected data is analysed using the cloud. The results presented states the reliability of the air quality sensing data to change behaviour patterns to promote health and well-being.

A cost-effective urban $PM_{2.5}$ supervision system which incorporates mobile sensing features is proposed by [26]. The proposed method incorporates a humidity and temperature sensor, GPS, a dust sensor and a noise sensor. The tests conducted are favourable and can ensure the ability to supervisor air quality at low-cost with reasonable precision.

Air quality sensing is an essential requirement for smart cities and further for global public health. Consequently, the design of cost-effective supervision solutions is a trending and relevant research field. Concerning the quality and appropriate contribution of various actual monitoring solutions [15, 23–26], a shortened comparison study is presented in Table 1.

Table 1. Summarised comparison review table of air quality sensing solutions.

MCU	Architecture	Sensors	Connectivity	Data consulting	Notifications	Ref.
PIC 24F16KA102	IoT	Temperature, humidity and CO_2	nRF24L01	Desktop and mobile	×	[23]
Arduino UNO	WSN	CO_2	ZigBee	×	×	[15]
Waspmote	WSN	CO, CO2, PM, temperature and relative humidity	ZigBee	Desktop	×	[24]
STM 32F103RC	IoT	PM, temperature and humidity	IEEE 802.15.4 k	Web and mobile	×	[25]
Mosaic-Node	StandAlone	GPS, PM2.5, noise, humidity and temperature	×	×	×	[26]

After the review of Table 1, the authors assume that the presented solutions incorporate different microcontroller units based on PIC, Arduino, Waspmote and STMicroelectronics. Moreover, various architectures are used in the development and design of air quality monitoring solutions. The proposed systems can be based on IoT, WSN or be standalone solutions. The most sensors for air quality monitoring are used to measure CO_2. One of the best indicators for air quality is the CO_2 concentration as it is emitted in high amounts and is comparatively effortless measured.

Regarding connectivity, the numerous communication technologies are used, and they are typically wireless. Several methods for data consulting are available such desktop, mobile and web applications. However, the solutions available in the literature does not incorporate notifications and easy installation methods. Notifications can be essential to advise the building or city manager to project arbitrations to improve the air quality on time.

In sum, the literature review presents several automatic systems and projects for air quality sensing, which support the relevance of the IoT architecture and mobile computing technologies as having a significant role in promoting health and well-being.

3   Air Quality Sensing for Enhanced Smart Cities

Smart homes are an essential and integral part of smart cities which had been studied for decades [27]. Smart homes are intended to meet several main applications such as the occupant's activity recognition to promote health and well-being, the use of multimedia technologies for monitoring, wherever the information obtained in the context is managed to trigger a notification for the occupant's protection. Additionally, the smart homes which the main propose is to improve energy efficiency using electric devices automation [28]. Numerous technological enhancements had decreased the cost of the implementation and design of smart homes. However, the development of automatic and intelligent systems for smart homes remains a significant challenge. Likewise, it is imperative to develop a system which can be easily installed and configured both in smart and traditional buildings. In the recent past, the number of sensors and actuators on the smart city context has increased, which lead to the creation of rich datasets for further investigation and data analytics. Smart homes present an effective method for enhanced living environments and improve numerous daily routine activities and provide the digitalisation of people's everyday activities [29]. In general, smart homes are developed for older adults to provide reliable healthcare systems and adequate realtime supervision of the occupant's health and well-being and to deliver feedback to the caregiver. An essential challenge for the smart city and smart home context is the security and privacy of the collected data. Therefore, high-quality research regarding the creation of secure methods to handle sensitive data is demanded [30, 31]. Air quality sensing must be assumed as significant in the smart city context, considering the air quality effects on people's health and well-being. Furthermore, indoor and outdoor air quality must be provided to improve people's health anytime and everywhere.

Nowadays, smartphones support excellent capabilities for data communication and processing, which can be assumed to consult the air quality data in real-time as people carry them in their daily lives. Moreover, buildings support numerous communication technologies and incorporate automation features. The vehicles are also crucial for the mobility of people in smart cities and are an essential source of air pollution. After combining air quality sensing for indoor and outdoor pollution monitoring, the city manager can achieve several benefits for public health and well-being. The gathered data can be used to project arbitrations on the traffic to decrease the pollution levels in the most critical locations of the city.

Additionally, the smart homes must incorporate air quality sensing to detect in useful time unhealthy situations and plan interventions to avoid them to occur. The air quality data must be validated before been stored and should be accessible anytime, anywhere. Furthermore, the data consulting method should be pervasive and ubiquitous and provide enhanced visualisation tools for data analytics. Specialists can analyse this information to assistance patient diagnostics and associate well-being complications with exposure conditions.

Air quality sensing includes several different stages such as data collection, data processing, data imputation, data storage and data analytics to deliver real-time notifications and alerts for enhanced public health (Fig. 1).

Moreover, the air quality data analysis can trigger critical alarms and notifications for the city and building managers using middleware services. The data collected by

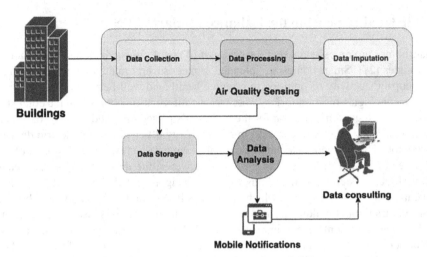

Fig. 1. Schema of an air quality sensing roadmap for enhanced living environments on smart cities.

the air quality monitoring systems are examined before storage. If this data surpasses the established thresholds, a warning should be triggered to support the occupants and city managers to act on time to promote air quality by applying proper air purifying methods inside buildings or by providing different strategies for traffic flow. These real-time alerts offer numerous advantages to achieve various and effective changes because these notifications not only provide measures to improve air quality but also allow the recognition of air quality patterns while repeated harmful situations are identified and to plan interventions to stop them from occurring.

In sum, for all the stated reasons in this section, air quality sensing must be seen as an essential element of people's everyday routine and be incorporated in smart cities to promote health and well-being. This data can be evaluated by professionals to achieve several health benefits that in another way, are impossible to address. Air quality monitoring allows the creation of rich datasets with spatiotemporal behaviour data, which will enable data consulting for city administrators to facilitate accelerated and effective arbitrations to promote public health.

4   Internet of Things Architecture for CO_2 Monitoring System: A Case Study

Taking into account all the requirements of air quality sensing, the authors develop the *MoniCO_2* solution. In this section, the detailed materials and methods used in the design of the proposed system are presented.

The *MoniCO_2* is an air quality supervision solution, which incorporates easy installation and configuration. The proposed system uses an ESP8266 which provides built-in wireless networking technologies capabilities and includes an SGP30 Multi-Pixel Gas sensor as a sensing unit. The central goal of the recommended method is to promote health

and well-being in smart cities. The proposed method wishes to deliver a mobile computing solution for air quality management to improve public health and safety (Fig. 2). The
authors develop a cyber-physical system based on the ESP8266 module, which supports
the IEEE 802.11 b/g/n protocol. The data collected is stored in a Microsoft SQL Server
database using ASP.NET web services. Mobile computing software has been developed
for data consulting using the Swift programming language.

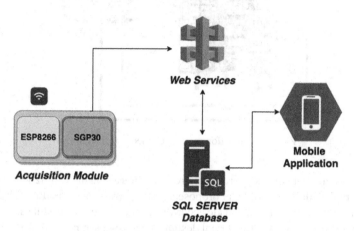

Fig. 2. *MoniCO$_2$* system architecture.

Furthermore, the proposed data acquisition module has been developed using on
open-source frameworks, and is a cos-effective system, with numerous benefits when
associated with actual methods. The acquisition module uses a FireBeetle ESP8266
(*DFRobot*) as a microcontroller, and the SGP30 gas sensor module (*Adafruit*) is
connected using the I2C interface. Figure 3 presents the prototype designed by the
authors.

The FireBeetle ESP8266 is a wireless board with unified antenna switches, power
and low noise amplifiers which is compatible with 802.11 b/g/n protocols. Additionally,
this board is WPA/WPA2 compatible and includes a 32-bit MCU and 10-bit ADC. The
selected board includes a 16 MB SPI flash memory.

The sensing unit is composed by the SGP30 Multi-Pixel Gas sensor, a metal-oxide
gas sensor developed at Sensirion. This sensor provides a calibrated output with 15% of
accuracy and provides an I2C connection for data communication [32]. The sensor range
for eCO_2 and TVOC concentration is 400–60,000 ppm and 0 to 60,000 ppb, respectively.
The eCO_2 output is based on a hydrogen measurement. The sensor sampling rate is 1 s,
and the average current consumption is 48 mA.

The proposed method offers a history of the variations of the indoor situation to
support the home or city administrator to deliver an accurate examination of the environment. The data collected can likewise to be employed to assist decision-making on
potential arbitrations for enhanced public health and well-being. The acquisition module
firmware is executed employing the Arduino Core, which is an open-source framework

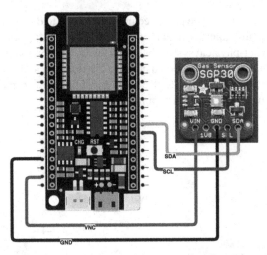

Fig. 3. Acquisition module connection diagram.

that intends to facilitate the use of regular Arduino libraries on the ESP8266 board without an external Arduino board. The proposed system cost is presented in Table 2. The choice of the sensor unit has been made according to the cost of the module as the principal intention was to study the functional design of the proposed method. The **MoniCO$_2$** is a low-cost system for air quality management, which costs an estimated 38.04 USD (Table 1).

Table 2. Amount value of the components used to design the proposed method.

Component	Amount (USD)
FireBeetle ESP8266	7.50
Adafruit SGP30 module	19.95
Cables and box	10.59
Overall cost	38.04

The proposed system offers a relevant function that delivers an easy configuration of the wireless network. The system is configured as a Wi-Fi client however if it is incapable of connecting to the wireless network or if no Wi-Fi networks are ready, the proposed system turn to hotspot mode and start a wireless network. At this time, this hotspot can be used to configure the Wi-Fi network to which the system will be connected through the introduction of the network access credentials.

The iOS application, denominated **MoniCO$_2$Mobile,** has been established using Swift in Xcode. The developed application has iOS 12 as minimum application requirements (Fig. 4). This mobile application includes numerous essential features, such as real-time data consulting using numeric tables and charts. Using the mobile application, people can inquire about the environmental data after authentication. Consequently, people can hold the parameters of chronicle records for additional examination. The introduced method is a decision-making tool to project arbitrations sustained by the data obtained for enhanced public health.

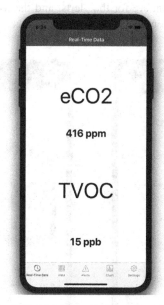

Fig. 4. *MoniCO₂Mobile* application: last collected data.

5   Current Status and Discussion

The proposed system was installed and tested inside a controlled laboratory environment. The system was tested continually for two months, and several experiments had been conducted with induced simulations. The module is powered using a 230 V-5 V AC-DC 2A power supply. The experiments indicate that the proposed air quality monitoring system can be used to detect unhealthy scenarios based on the CO_2 and TVOC data at low-cost. The data is collected every 30 s, but this value can be changed according to the needs of the installation. The *MoniCO₂Mobile* offers information checking as graphical or statistical modes. An example of the data gathered by the system is represented in Fig. 5. The graphs present the results gathered in the physical environment, including provoked simulations of smoke. The results are promising the proposed IoT architecture can provide accurate CO_2 monitoring in real-time and demonstrate a fast response on significative CO_2 variation.

The chart view available on the mobile application provides an improved inspection of observed parameters behaviours when compared with the statistical presentation. First, mobile software presents quick and straightforward access to the obtained data and allows further accurate investigation of air quality progression. Consequently, this method is a relevant instrument for air quality surveillance and to assist decision making on potential interventions to improve performance but also to detect exposure risk scenarios. Second, this IoT architecture offers chronological data evolution for improved air quality assessment, which is particularly appropriate to identify unhealthy scenarios and project interferences to improve health and well-being. Moreover, the mobile application provides a configuration are for the alerts, and the user can easily change that values according to the requirement of the monitored space.

Fig. 5. *MoniCO₂Mobile* application: settings and chart view.

The mobile software provides fast and intuitive access to the collected data as mobile phones have today high performance and storage capacities, and most people use them every day. In this way, the building or city manager can transport the air quality data of their environment for further analysis.

Currently, most air quality monitoring systems are high-priced and are established on random sampling. Therefore, these methods are limited by offering data associated with a particular sample without temporal association. Right to be told, the professional systems available on the market are accurate and developed for industrial activities. Further, some solutions are portable and compact. However, these solutions do not support real-time information accessibility for city authorities or build managers. These real-time alerts are essential to facilitate accelerated and decisive interference to promote health and safety. The existing industrial products offer data history constricted to the system memory and required special software for data downloading and administration. The design and development of innovative air quality sensing systems based on the literature which support real-time data consulting for examination and visualisation must be assumed as an essential enhancement regarding the smart city objectives.

Compared with the systems presented in Table 1, the proposed solution provides adaptability and precision of analysis in real-time, providing a meaningful progression of the actual air quality monitoring systems. The results are promising as the proposed system can be used to provide a correct air quality assessment at low-cost. The proposed method can be used to support city managers to detect possible unhealthy scenarios in smart cities in useful time. Another essential benefit of the presented method is the scalability related with the modularity of the solution. The installation can be done through one module, but other modules can be added in the future regarding the requirements of the environment.

IoT is a standard that intends to increase the quality of personal experience. IoT architectures and AAL will reciprocally provide systematic improvements leading to the design of automated and intelligent systems to promote health and well-being. IoT is as a meaningful concept to improve daily routine actions in numerous fields such as buildings, manufacturing, healthcare, smart homes and smart cities. Shortly, innovative methods will improve the number of intelligent solutions established in traditional people houses.

The authors consider that soon, most of the residences will be controlled in real-time and implement warnings to inform the homeowner in the occurrence of inadequate ambient quality. Our living environments will support monitoring features, and the data gathered will be distributed with the therapeutic specialist to compare wellness complications among inadequate practices of the patients and assist clinical assessment.

6 Conclusion

The smart city concept is an approach to embrace current urban production in a mutual context as cities face relevant open issues regarding the socio-economic development requirements and to improve people well-being. IoT will be able to offer the operability between different systems to develop a centralised urban-scale framework for enhanced smart cities.

Smart homes are an essential and integral part of smart cities and are intended to meet several applications to promote health and well-being. Most individuals stay most of their time indoors, and therefore, smart homes are unquestionably an extraordinary place to monitor. Moreover, IoT architectures can provide ubiquitous and pervasive methods for environmental data acquisition and provide connectivity for data transmission. Particularly, IAQ has a material adverse impact on occupational health and well-being. Therefore, air quality sensing must be seen as an integral part of society's everyday activities and must be incorporated in smart cities to promote health and well-being.

Moreover, this data can be evaluated by health professionals for diagnostics support and to correlate patient problems with their living environment. Nevertheless, the air monitoring solutions available in the literature does not incorporate notifications and easy installation methods. Notifications can be essential to advise the building or city manager in real-time to design mediations to promote the air quality in useful time. Therefore, this manuscript had performed a review and roadmap for air quality sensing and proposed an IoT architecture for air quality supervision. The results are encouraging as the suggested system can be used to provide proper air quality assessment. The proposed method can be used to support city managers to detect possible unhealthy scenarios in smart cities early. Furthermore, mobile software provides fast and intuitive access to the collected data, as actually, most people use mobile phones.

References

1. Giusto, D. (ed.): The Internet of Things: 20th Tyrrhenian Workshop on Digital Communications. Springer, New York (2010). https://doi.org/10.1007/978-1-4419-1674-7
2. Marques, G., Pitarma, R., Garcia, N.M., Pombo, N.: Things architectures, technologies, applications, challenges, and future directions for enhanced living environments and healthcare systems: a review. Electronics **8**, 1081 (2019). https://doi.org/10.3390/electronics8101081
3. Atzori, L., Iera, A., Morabito, G.: The Internet of Things: a survey. Comput. Netw. **54**, 2787–2805 (2010). https://doi.org/10.1016/j.comnet.2010.05.010
4. Marques, G.: Ambient assisted living and Internet of Things. In: Cardoso, P.J.S., Monteiro, J., Semião, J., Rodrigues, J.M.F. (eds.) Harnessing the Internet of Everything (IoE) for Accelerated Innovation Opportunities, pp. 100–115. IGI Global, Hershey (2019). https://doi.org/10.4018/978-1-5225-7332-6.ch005
5. Caragliu, A., Del Bo, C., Nijkamp, P.: Smart cities in Europe. J. Urban Technol. **18**, 65–82 (2011). https://doi.org/10.1080/10630732.2011.601117
6. Schaffers, H., Komninos, N., Pallot, M., Trousse, B., Nilsson, M., Oliveira, A.: Smart cities and the future internet: towards cooperation frameworks for open innovation. In: Domingue, J., et al. (eds.) FIA 2011. LNCS, vol. 6656, pp. 431–446. Springer, Heidelberg (2011). https://doi.org/10.1007/978-3-642-20898-0_31
7. Chourabi, H., et al.: Understanding Smart Cities: An Integrative Framework. Presented at the January (2012). https://doi.org/10.1109/HICSS.2012.615
8. Zanella, A., Bui, N., Castellani, A., Vangelista, L., Zorzi, M.: Internet of Things for smart cities. IEEE Internet Things J. **1**, 22–32 (2014). https://doi.org/10.1109/JIOT.2014.2306328
9. Batty, M., et al.: Smart cities of the future. Eur. Phys. J. Spec. Top. **214**, 481–518 (2012). https://doi.org/10.1140/epjst/e2012-01703-3
10. Hernández-Muñoz, J.M., et al.: Smart cities at the forefront of the future internet. In: Domingue, J., et al. (eds.) FIA 2011. LNCS, vol. 6656, pp. 447–462. Springer, Heidelberg (2011). https://doi.org/10.1007/978-3-642-20898-0_32
11. Rashidi, P., Mihailidis, A.: A survey on ambient-assisted living tools for older adults. Biomed. Health Inform. IEEE J. Of. **17**, 579–590 (2013). https://doi.org/10.1109/JBHI.2012.2234129
12. Seguel, J.M., Merrill, R., Seguel, D., Campagna, A.C.: Indoor air quality. Am. J. Lifestyle Med. (2016). https://doi.org/10.1177/1559827616653343
13. Bruce, N., Perez-Padilla, R., Albalak, R.: Indoor air pollution in developing countries: a major environmental and public health challenge. Bull. World Health Organ. **78**, 1078–1092 (2000)
14. Myers, S.S., et al.: Increasing CO_2 threatens human nutrition. Nature **510**, 139–142 (2014)
15. Salamone, F., Belussi, L., Danza, L., Galanos, T., Ghellere, M., Meroni, I.: Design and development of a nearable wireless system to control indoor air quality and indoor lighting quality. Sensors **17**, 1021 (2017). https://doi.org/10.3390/s17051021
16. Akkaya, K., Guvenc, I., Aygun, R., Pala, N., Kadri, A.: IoT-based occupancy monitoring techniques for energy-efficient smart buildings. In: 2015 IEEE Wireless Communications and Networking Conference Workshops (WCNCW), pp. 58–63. IEEE, New Orleans (2015). https://doi.org/10.1109/WCNCW.2015.7122529
17. Marques, G., Pitarma, R.: A cost-effective air quality supervision solution for enhanced living environments through the Internet of Things. Electronics **8**, 170 (2019). https://doi.org/10.3390/electronics8020170

18. Marques, G., Ferreira, C.R., Pitarma, R.: Indoor air quality assessment using a CO_2 monitoring system based on Internet of Things. J. Med. Syst. **43** (2019). https://doi.org/10.1007/s10916-019-1184-x

19. Marques, G.M.S., Pitarma, R.: Smartphone application for enhanced indoor health environments. J. Inf. Syst. Eng. Manag. **1** (2016). https://doi.org/10.20897/lectito.201649

20. Marques, G., Pitarma, R.: Monitoring and control of the indoor environment. In: 2017 12th Iberian Conference on Information Systems and Technologies (CISTI), pp. 1–6. IEEE, Lisbon, Portugal (2017). https://doi.org/10.23919/CISTI.2017.7975737

21. Marques, G., Pitarma, R.: IAQ evaluation using an IoT CO_2 monitoring system for enhanced living environments. In: Rocha, Á., Adeli, H., Reis, L.P., Costanzo, S. (eds.) WorldCIST 2018. AISC, vol. 746, pp. 1169–1177. Springer, Cham (2018). https://doi.org/10.1007/978-3-319-77712-2_112

22. Marques, G., Pitarma, R.: An indoor monitoring system for ambient assisted living based on Internet of Things architecture. Int. J. Environ. Res. Public. Health. **13**, 1152 (2016). https://doi.org/10.3390/ijerph13111152

23. Shah, J., Mishra, B.: IoT enabled environmental monitoring system for smart cities. In: 2016 International Conference on Internet of Things and Applications (IOTA). pp. 383–388. IEEE, Pune, India (2016). https://doi.org/10.1109/IOTA.2016.7562757

24. Bhattacharya, S., Sridevi, S., Pitchiah, R.: Indoor air quality monitoring using wireless sensor network. Presented at the December (2012). https://doi.org/10.1109/ICSensT.2012.6461713

25. Zheng, K., Zhao, S., Yang, Z., Xiong, X., Xiang, W.: Design and implementation of LPWA-based air quality monitoring system. IEEE Access **4**, 3238–3245 (2016). https://doi.org/10.1109/ACCESS.2016.2582153

26. Gao, Y., et al.: Mosaic: a low-cost mobile sensing system for urban air quality monitoring. In: IEEE INFOCOM 2016 - The 35th Annual IEEE International Conference on Computer Communications, pp. 1–9. IEEE, San Francisco, CA, USA (2016). https://doi.org/10.1109/INFOCOM.2016.7524478

27. Moukas, A., Zacharia, G., Guttman, R., Maes, P.: Agent-mediated electronic commerce: an MIT media laboratory perspective. Int. J. Electron. Commer. **4**, 5–21 (2000)

28. De Silva, L.C., Morikawa, C., Petra, I.M.: State of the art of smart homes. Eng. Appl. Artif. Intell. **25**, 1313–1321 (2012). https://doi.org/10.1016/j.engappai.2012.05.002

29. Wilson, C., Hargreaves, T., Hauxwell-Baldwin, R.: Smart homes and their users: a systematic analysis and key challenges. Pers. Ubiquit. Comput. **19**, 463–476 (2015)

30. Shen, J., Wang, C., Li, T., Chen, X., Huang, X., Zhan, Z.-H.: Secure data uploading scheme for a smart home system. Inf. Sci. **453**, 186–197 (2018). https://doi.org/10.1016/j.ins.2018.04.048

31. Dorri, A., Kanhere, S.S., Jurdak, R., Gauravaram, P.: Blockchain for IoT security and privacy: the case study of a smart home. In: 2017 IEEE International Conference on Pervasive Computing and Communications Workshops (PerCom Workshops), pp. 618–623. IEEE, Kona, Big Island, HI, USA (2017). https://doi.org/10.1109/PERCOMW.2017.7917634

32. Rüffer, D., Hoehne, F., Bühler, J.: New digital Metal-Oxide (MOx) sensor platform. Sensors **18**, 1052 (2018). https://doi.org/10.3390/s18041052

Citizen-Centre Needs

Smart Pedestrian Network: An Integrated Conceptual Model for Improving Walkability

Fernando Fonseca[1](✉) ⓘ, Paulo Ribeiro[1] ⓘ, Mona Jabbari[1] ⓘ, Elena Petrova[2] ⓘ,
George Papageorgiou[3] ⓘ, Elisa Conticelli[4] ⓘ, Simona Tondelli[4] ⓘ, and Rui Ramos[1] ⓘ

[1] Centre for Territory Environment and Construction (CTAC),
University of Minho, Guimarães, Portugal
ffonseka@gmail.com
[2] Association for Sustainable Innovative Development in Economics, Environment and Society
(ASIDEES), Vienna, Austria
[3] European University Cyprus, E.U.C. Research Centre, Nicosia, Cyprus
[4] Alma Mater Studiorum, University of Bologna, Bologna, Italy

Abstract. Smart and sustainable mobility have recently emerged as a solution to the problems incurred by the intensive use of motorised transport modes. For many decades, cities have been planned based on the needs of vehicle traffic, neglecting basic human needs for active mobility and the adverse effects of motorised traffic on the natural environment. However, walking is an environmentally friendly transport mode and a healthy form of making physical activity. Thus, walking becomes an essential component of the transport and urban policies for achieving a more sustainable development process. This paper presents the research project Smart Pedestrian Network (SPN) that aims at promoting walkability as one of the critical dimensions of smart and sustainable mobility in cities. The paper analyses the various components linked to SPN that can make a pedestrian network "smart" and, therefore, a feasible alternative to motorised transport modes. Three integrated components are analysed: (i) an urban planning component supported in a GIS-based multi-criteria model to assess the conditions provided to pedestrians and to support the adoption of planning policies; (ii) a smartphone app for pedestrian navigation, displaying optional routes according to the pedestrian preferences and needs; and (iii) a business component to estimate and disseminate the multiple benefits of walking as well as the market potential of SPN. By promoting an innovative linkage of these three components, SPN has a great potential for improving walkability and, therefore, for creating more sustainable and liveable urban spaces.

Keywords: Smart Pedestrian Network · Walkability · Urban planning · Smartphone app · Walking benefits

1 Introduction

It can certainly be said that the last century was the century of cars. Private motorised vehicles nurtured the sense of individual freedom and radically changed the way cities

P. Pereira et al. (Eds.): SC4Life 2019, LNICST 318, pp. 125–142, 2020.
https://doi.org/10.1007/978-3-030-45293-3_10

have been designed and conceived. Throughout the past century, cars colonised the spaces of everyday life, generating visible effects in cities around the world such as public spaces invaded by parked vehicles, large streets serving huge suburbs, giving birth to an uncontrolled urban expansion that consequently generated situations of social isolation and segregation. As highlighted by Jacobs [1] cities are no longer conceived as a set of spaces and buildings, but only as individual buildings. High motorisation rates in conjunction with functionalism theories increased the separation of urban activities and spaces, turning the cities more fragmented and car-dependent. Nowadays, in the European cities, the private car is intensively used including for short urban trips: about 30% of the car trips are lower than 3 km, while 50% are lower than 5 km [2]. The intensive use of the car on urban trips is not only a matter of space and distance. It also produces environmental and health impacts. In European cities, 40% of CO_2 emissions and up to 70% of other pollutants are related within the intensive use of cars [3]. Noise pollution, consumption of non-renewable resources are other environmental consequences related to the use of the car. In the health domain, the use of car contributes to physical inactivity, which is a leading risk factor for premature mortality and various health problems associated to sedentary lifestyles, such as obesity, diabetes and depression [4].

In Europe, 73% of the population lives in cities, and it is expected that this number will rise to 80% by 2050 [5]. The concentration of the population in cities is causing several environmental, spatial, economic and social problems. One of the main problems is related to the intensive use of private cars, namely for short urban trips. In Europe, about 30% of the urban car trips are lower than 3 km, while 50% are lower than 5 km [2]. Note that a viable alternative transport mode for such short distances could simply be walking.

The intensive use of motorised vehicles causes several well-known problems, including pollution, traffic congestion, energy consumption, accidents and sedentary lifestyles [6]. In European cities, 40% of CO_2 emissions and up to 70% of other pollutants are related within the intensive use of cars [3]. By 2050, the EU has the goal of cutting transport emissions to 60% below 1990 levels to make cities more sustainable and liveable [7]. It is recognised that effective urban and transport planning policies should mitigate these problems namely by developing disruptive mobility solutions [8] so that the United Nations' Sustainable Development Goals can be achieved [9]. To make the transition to a low-carbon society, a new mobility paradigm, less supported on private car, is necessary [10]. The concepts of smart mobility and smart city have recently emerged to overcome these problems. To be considered smart, mobility must be sustainable [11]. The sustainable mobility concept encourages a modal shift towards more sustainable forms of transport namely by increasing the use of active modes such as cycling and walking [12]. The benefits of adopting more sustainable modes of transport are generally appreciated by people, but the various approaches to shift mobility paradigm failed due to the complex factors that prevent people for adopting behavioural change [13].

Walking is an environmentally friendly transport mode and a viable alternative mode of transport for short urban trips. Walking has many advantages when compared with car trips. In the environmental domain, walking consumes less non-renewable resources, reduces the emission of air pollutants, CO_2, and noise [14, 15]. In the economic domain, walking is costing far less than motorised transports, both in direct user costs and public infrastructure costs. Walking reduces health care costs, improves work productivity and has high accessibility and low complexity [14]. As the only energy required is provided by the traveller, walking also presents several health outcomes in terms of reducing obesity, cardiovascular diseases, diabetes, stress, among others [13, 15]. For all these benefits, walking gained enormous popularity for its potential to promote more sustainable mobility and healthier lifestyles.

In this context, the goal of this paper is to present the research project "SPN - Smart Pedestrian Net". SPN is a pilot innovative European research project to assess and improve the walkable conditions provided to pedestrians. The overall goal is to develop a model to support European cities to become people-oriented by improving walkability as one of the important dimensions of smart, sustainable and inclusive cities. The paper explains the concept of the Smart Pedestrian Network and how the multiple dimensions attached to it in the planning, technological and business domains were considered.

2 Literature Review

Pedestrian studies can be divided into two main groups: (i) studies focused on assessing the conditions provided to pedestrians; and (ii) studies focused on analysing pedestrians behaviours and preferences. The first group integrates the studies on walkability, a concept widely used that reflects the quality of the walking conditions or, in other words, the extent to which the built environment supports and encourages walking safety [16]. Walkability is a multidimensional concept that combines several attributes or criteria with an impact on walking [17]. These criteria traditionally rely on 3Ds (density, diversity, and design) features of the built environment proposed by Cervero and Kockelman [18] plus the 2Ds (distance to transit and destination accessibility) added later by Cervero et al. [19]. Thus, physical environment mesoscale attributes such as land use mix, proximity to destinations, density and connectivity have been widely used in walkability studies [20, 21]. Microscale walkability approaches at the level of streetscapes have also been developed for evaluating walkability [22–24]. Objective measures have been predominantly used in walkability research by developing walkability audits [25, 26] and indexes [27–29] that evaluate and score the conditions provided to pedestrians. Such objective measures have been supported in Geographic Information Systems [30, 31], multi-criteria analysis [25, 32], virtual technologies [29, 33], Web-based services, such as WalkScore, among others. Subjective measures have been supported on qualitative assessments based on stated preferences and individual perceptions usually collected through questionnaires or by consulting expert panels [34].

Walkability has been analysed by assessing different criteria. Such criteria can be grouped in the following dimensions: accessibility, safety/security, pedestrian facilities, and land use. Accessibility is in the first level of walking needs [35] and can be described as the ability to reach desired destinations, making an acceptable effort [26]. Accessibility has been often quantified as the distance to public transport [36] and key amenities, such as schools , shops and urban parks [37]. It is widely recognised that 400 and 800 m are the referential walking distance to get bus stops and train stations, respectively [38]. In turn, safety refers to pedestrians being protected from traffic. Traffic safety is critical for making walkable environments and plays a significant role in the decision to walk [39], particularly for children and seniors. The characteristics of traffic (speed, volume), the characteristics of roads (lanes and directions) and the facilities for protecting pedestrians from traffic are factors with impact on walking. Public security is also a criterion with some impact on walking, mainly in cities characterised by urban violence [32]. Pedestrian facilities are infrastructure provided to enhance the pedestrian environment, to improve pedestrian mobility, safety, access, and comfort. For that reason, pedestrian facilities have been one of the most analysed factors, namely by considering the following criteria: the characteristics, maintenance and continuity of sidewalks [28], the presence of obstacles on sidewalks that creates discomfort and affects pedestrian safety [40]; slopes, as small positive increments in slopes decrease travel speeds and increase the energy and the effort required for walking [41, 42]; the presence of trees and greenery that bring several environmental, safety and aesthetic benefits and encourage people to walk and the walking experience [25]; and street furniture that helps in creating a more pleasant and attractive walkable environment [43]. Finally, the characteristics of land use have been found to affect mode choice and walking [44]. Land use is often operationalised using density, diversity and design measures. Areas with high densities (population, residential, commercial) usually generate more people walking, by decreasing the appeal of driving through congested areas where parking is often scarce [25, 45]. Densities are usually analysed by developing ratios and indexes. Diversity expresses the degree to which there is a mix of land uses within an area [46]. Mixed land use is found to be one of the most correlated characteristics with utilitarian walking [42, 45]. Entropy and dissimilarity indexes have been adopted to evaluate the level of diversity [26]. Urban design and planning practices also have a determinant impact on walking [22, 24, 39]. Some authors, such as Ewing and Handy [47] and Ewing et al. [22], developed tools and specific methods for measure urban design.

On the other hand, behavioural researches have been mostly focused on pedestrian choices and preferences [48]. As pedestrians move more slowly than vehicles, they are more sensitive to their surroundings, and numerous aspects affect the experience and behaviour of pedestrians [49]. These include aspect related to the neighbourhood environment (the meso and microscale criteria mentioned above), as well as individual factors, including demographic, socioeconomic and psychological aspects, such as the age, gender, and income. In this topic, some authors have studied how the individual characteristics of pedestrians and their physical condition influence their preferences and behaviours. For example, Bernhoft and Carstensen [50] demonstrated that older pedestrians prefer routes with crossings and signalised intersections more than young pedestrians. Giles-Corti et al. [39] also found that crossing major streets and areas with high intersection density were negatively associated with child's active travel to school. Walking behaviours are also affected by gender. Many studies on trips to school have found that girls are less likely to walk than boys [51]. Among university students, Delmelle and Delmelle [52] also found that females are less like to walk (and bike) to university than male students . Female students tend to be more cautious to the risk of active travelling on streets shared with traffic [52] and are also generally more concerned with security especially after dark [53]. Male students are found to be more likely to switch commuting modes throughout the year, while females are generally more likely to drive [52]. In turn, Moniruzzaman and Farber [54] also found that students from families with higher incomes are less available to use active modes and prefer to drive instead of walk. These walking behavioural approaches have been mostly carried through questionnaires and surveys [39, 51, 52] but also by using advanced spatial simulation tools [55] and mobile devices [24].

In sum, the literature shows that the option to walk and the choice of a specific route depend on a variety of different criteria. Individuals' decisions about where and whether to walk are highly complex and typically entail considering multiple factors, including internal factors related to the personal characteristics of individuals, as well as external factors related to the characteristics of the environment [38]. Distance and time to walk, perceived ease, comfort, safety, security, convenience, proximity and attractiveness of the route, as well as age, gender, ethnicity, income, household size, car ownership, among others are factors that influence the decision to walk and which route to take.

3 Method for a Smart Pedestrian Network

The concept of a Smart Pedestrian Network (SPN) aims at developing more walkable cities and increase the number of people who walk. The goal of SPN consists in developing an integrated concept to analyse the various aspects that can make a pedestrian network "smart" and therefore, a feasible alternative to motorised transport modes. SPN sustains that a "smart pedestrian network" is not reduced to a sustainable dimension resulting from replacing the car by walking, neither to the concept of an interconnected set of appropriate sidewalks covering a given urban area. The project sustains that implementing a Smart Pedestrian Network requires a broader and multi-linked integrated approach to support not only the adoption of efficient planning initiatives but also to provide data to pedestrians about the environmental, economic and health benefits of walking, by using the potential given by digital services and devices. Approaches only focused on specific issues can have a much-limited impact. For instance, the development of navigation tools for pedestrians only makes sense if the city provides minimum conditions for walking. Otherwise, people will not feel comfortable and safe and will prefer to drive instead of walk. Thus, a smart approach requires the development of innovative technological tools to help pedestrians in selecting routes according to their preferences as well as the dissemination of the benefits (externalities) of walking. Both components are vital for changing the paradigm of mobility, and for supporting policies and investments that could have a real impact on modal choice. In the following subsections, the roadmap to the smart pedestrian network is shown in Fig. 1, considering three dimensions: urban planning, smartphone app and externalities/business model.

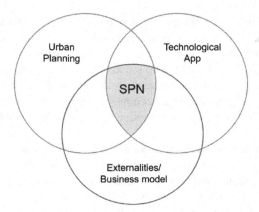

Fig. 1. Proposed roadmap for a Smart Pedestrian Network.

3.1 Roadmap for a Smart Pedestrian Network: Urban Planning

In the urban planning domain, the goal of SPN is to provide a tool to evaluate the conditions provided to pedestrians, to identify the areas more and less walkable and, therefore, to provide urban and transport guidelines for improving walkability and sustainable mobility. The core of the proposed tool is a GIS-based multi-criteria model, combining supply and demand-based walking fundamentals. The structure of the SPN model is presented in Fig. 2. Subsequent paragraphs, however, are indented.

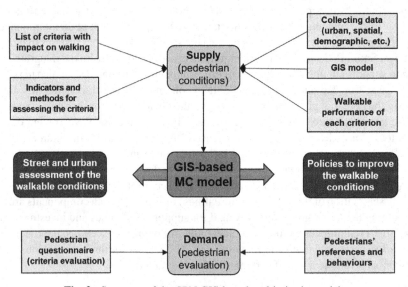

Fig. 2. Structure of the SPN GIS-based multicriteria model.

As shown in the literature review, walkability is related to multiple interconnected criteria. Multi-criteria analysis is considered an appropriate tool for dealing with a wide range of criteria, for solving complex problems by assessing multiple solutions, yielding results that are more effective, clear and logical than single-criteria approaches, for helping decision-makers in finding solutions [56]. For that reason, multi-criteria analysis becomes a widely used decision-making technique in the urban and transport domains [57].

On the supply side and for developing the multi-criteria analysis, the first step consists in defining an extensive list of interdependent criteria with an impact on walking. As emphasised by Lin and Wei [58], many authors have based their assessments on a limited and subjective number of criteria. To overcome this debility, SPN was supported in an extensive list of 23 micro and mesoscale criteria, divided by the following four dimensions: accessibility, safety and security, pedestrian facilities, land use and urban design. The list of criteria is presented in Table 1.

Table 1. Dimensions and criteria included in the SPN model.

Dimensions	Criteria
Accessibility	Distance to bus stop
	Distance to light rail stations
	Distance to train stations
	Distance to public amenities
	Distance to car parking
Street connectivity	Intersection density
	Street integration
Safety and security	Traffic speed (km/h)
	Traffic lanes (number)
	Pedestrian crossings
	Public security (street light, graffiti, abandoned buildings)
Pedestrian facilities	Width of sidewalks
	Condition of sidewalks
	Obstacles on sidewalks
	Street furniture
	Slopes (%)
	Trees and greenery
Land use	Population density
	Residential density
	Non-residential density
	Land use mix
Urban design	Human scale/enclosure
	Complexity
	Transparency

For each criterion shown in Table 1, the model specifies the respective indicators and how each criterion should be assessed. Indicators were depicted from literature review specifying, for instance, the distance from which transport stations are within a walkable distance, defining ratios for analysing densities, describing which slopes are more comfortable for pedestrians, minimum sidewalks width, etc. The assessment involves both quantitative and qualitative assessment that requires detailed spatial data, such as disaggregated census data at the tract level, street and traffic data, digital models of the terrain, among others. This data is also required for developing the GIS-based model, where each criterion should be represented by a specific shapefile (GIS file). In a GIS environment is possible to visualise, analyse and map the performance of each criterion within a specific study area, as well as to make broader walkable analyses by overlapping several criteria.

After defining the criteria and developing the GIS model, the next step consists in integrating the demand side on the model by performing an extensive questionnaire to pedestrians. The goal of the pedestrian questionnaire is to include the pedestrians' perception on the model (a people-oriented model), namely to (i) define the relative importance of criteria (weighting process); and (ii) to collect additional mobility data (routing preferences, use of mobile devices, pedestrian behaviour, etc.). Converting the pedestrians' evaluation into values require operations of fuzzy logic, linear normalisation and analytic hierarchy process. The result is an assessment reflecting not only the walkable conditions objectively assessed in the city but also the evaluation made by pedestrians. The assessment can be focused on a specific criterion or in a specific part of the network (microscale approach), as can be extended to analyse the conditions provided by multiple/interdependent criteria or to the city level (macroscale approach). Based on the overall walkable levels assessed by the model, several planning measures can be proposed to improve the conditions provided to pedestrians. Planning guidelines should be discussed in close collaboration with city planners, decision-makers and other experts considering the pedestrians' needs and the problems identified.

3.2 Roadmap for a Smart Pedestrian Network: Smart Pedestrian Assistant Mobile Application as IT Solution

The concept of smart city is usually linked to the use of technological tools, which allow modern cities to enhance the quality of the services provided to citizens [59]. A wide range of public services can be deployed as part of smart city initiatives, including in the transport domain. For that reason, the project sustains that implementing a Smart Pedestrian Network also requires developing technological tools, such as pedestrian navigation tools, to aid people to walk and to show the benefits of walking. There are many navigations systems in the market but they mainly address the needs of vehicle users and do not take into account the conditions provided to pedestrians [60]. Thus, a pedestrian navigation system running on mobile devices, such as smartphones, are particularly useful for providing navigational aid to users at unfamiliar places and additional data related to their trips (tracing the routes, obtaining health data, etc.).

The mobile app Smart Pedestrian Assistant (SPA) developed in SPN introduced an innovative design and multiple new functions realising the key project concepts as follows:

- Identifying optional pedestrian routes between one specific origin and destination according to the pedestrian preferences and interests, including typical walks with kids, dogs, friends, invalids, in green areas, etc.;
- Providing three types of routes corresponding to the selected criteria (Fig. 3);
- Changing the offered route options over time avoid always the same routes and offer to pedestrians novel walking impressions;
- Taking into consideration passed route evaluations and statistics by pedestrians;
- Proving optional selection of flat routes, green routes, and safe routes (e.g. including benches, avoiding high traffic noisy streets, etc.);
- Inserting pedestrian data (age, gender, disability) during the routing process. For instance, elderly or disable pedestrians may prefer flat routes because they require less physical effort;
- Allowing pedestrians to express their physical and psychological satisfaction (visual aspects, comfort, security, tranquility) to improve the linkage between the urban and technological components of SPN;
- Allowing pedestrians to define preferences regarding total route time, distance, points of interest to be visited (e.g. 1 park, 2 museums, 1 restaurant, 2 shops, etc.) and obtain recommended optional routes that allow the walk within the defined restrictions (e.g. be back at home by 17:00 or in 4 h);
- Considering only circle route staring from home or hotel that shall satisfy multiple conditions in timing, points of interests, number of steps, type of streets, etc.;
- Having the route monitored during the walk to review it ongoing status such as current duration of the walk, places visited and those which are left to be visited;
- Being promptly informed by real-time notifications during the route and specific current location of the pedestrian regarding optional events such as accidents, gas and water leaks, terror acts, etc.;
- Obtaining the municipality recommendations to change the route if necessary;
- Interrupting the route at any time by some reason and get assistance in finding the shortest routes to necessary city services (toilet, hospital, police, etc.);
- Providing feedback to the municipality regarding the quality of routes, streets and walkability conditions.

Fig. 3. Smart Pedestrian Assistant prototype support for SPN routing.

The structure of the novel SPN technological solution enabling these functionalities is presented in Fig. 4. It consists of the latest innovation cloud web platform SCM - Smart City Monitor (https://smartcity.pharosnavigator.com) that continuously collects various urban static and dynamic (real-time) data from different urban sources and systems and transforms it into rich set of information services for the municipality management including city/districts/streets status for pedestrians, transportation, waste, energy, pollution, among others. As powerful new AI-driven digital transformation technology, the SCM allows applying all the relevant criteria that impact on walking: street network, topography, transport stations, location of amenities, and the overall data necessary for using the selected criteria. SCM works with multiple SPA of the pedestrians walking the city and supports them by providing real-time walkability data linked to OpenStreetMap (used as standard mapping service and database) as well as static geographic data provided in GIS format (from the SPN urban planning component).

The integrated real-time urban information system consisting of SCM and multiple SPAs used by pedestrians walking in the city can support the variety of services for the urban community (street temperature, local pollution, air quality, weather and traffic variability) and provide updates regarding various local events. The services, however, depend on the availability of relevant urban data, city sensor networks and connectivity between SPA and SCM (Wi-Fi, cellular 4G, 5G) as well as the city information operational centre, which can support such complex urban processes monitoring and events management.

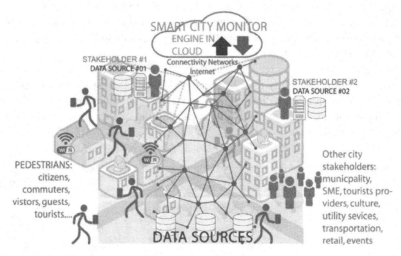

Fig. 4. Data sources and network objects for the SPN app.

3.3 Roadmap for a Smart Pedestrian Network: Externalities and Business Model

SPN provides information to pedestrians about the environmental, economic and health benefits of walking. The use of digital services and devices, such as SPA, has the potential for encouraging people to walk, including by providing data about the benefits of walking. Estimating the positive externalities of active travelling is a field not often analysed in the literature [61]. However, for SPN, a smart approach requires that providing data to people about the benefits of walking is vital for creating a paradigm shift in mobility, and for supporting policies and investments that could have a real impact on modal shift.

In this context, SPN also intends to estimate the benefits (externalities) of replacing the car by walking. Direct benefits for pedestrians include personal savings with fuel and car maintenance, travel time savings in short urban trips, health benefits resulting from walking, as well as benefits more difficult to quantify such as those resulting from social interaction with friends and family during group walks. Walking more and drive less also produce environmental benefits, namely in terms of reducing harmful emissions (CO_2, NOx, PMs, SO_2, noise) and reducing the consumption of non-renewable resources. Average travel distances, average speed and car emissions are variables that can be used for estimating these externalities. Some of these benefits are provided by the app namely those related to health (number of steps, calories burnt). The app can also be recalibrated for health recommendations namely for increasing the number of steps/day. The collaboration with some stakeholders is vital for estimating the many benefits that could potentially arise from SPN. For instance, in the case of the health benefits, working with insurance companies is critical to collect the necessary data to estimate the reduced healthcare costs incurred by insurance companies, governments and people.

A second component is related to the development of the business model for the SPN app. A business model is often described as the rationale of how an organisation creates, delivers and captures value [59]. Thus, a business model provides a coherent way to consider their options in uncertain, fast-moving and unpredictable environments. The SPN business plan [62] and model [63] is supported in a cost-benefit analysis for different scenarios related to the app. As shown in Fig. 5, the business model is divided into two phases. The first is related to the technological development of the tool and includes the following steps:

- Funding: secure funding and commitment for implementing SPA. Consult governments, municipalities other stakeholders, etc.;
- Research and design: attain the necessary tools, equipment and technology to begin the implementation of SPA;
- Testing and adapting: test the SPA with the participation of municipal authorities and end-users;
- Feedback: collect feedback from participants including pros and cons from their experience with SPN;
- Adjustments: make any necessary changes and adjustments to SPN to maximise its potential.

The second is more linked to marketing issues and includes the following steps:

- Marketing: with the expertise of the marketing team of SPN, the cities with the highest potential benefit of walking can be targeted.
- Launch: implement SPN in various European municipalities.

- Observe: collect data on SPN implementation, performance and feedback. Feedback will be collected continuously.
- Adjustments: make any necessary additional changes to SPN depending on user needs, geographical location, demographics etc.
- Global markets: once the team have enough experience with the European Market, the next target is the global market.

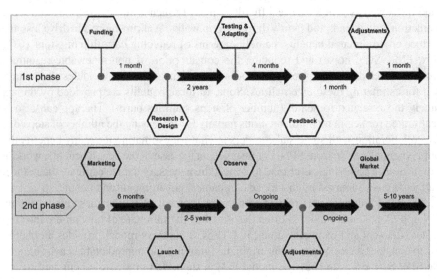

Fig. 5. Phases and steps of the SPN business model.

For each city of SPN prototype implementation, the SPN business model considers three scenarios: a Worst Case (1% users), an Average case (3% users), and a Best case (5% users). For each scenario, the business model takes into account incomes, as well as direct and indirect expenses for the premium version of SPA. Considering the multiple benefits of walking, there is an enormous potential for creating a truly disruptive innovative revenue model by buying and selling steps. Thus, the SPA can create measurable value for multiple stakeholders, including not only the SPA users (segmented customers) but also municipalities, governments, companies, among many others. As the navigation systems already available in the market do not include the quality of the pedestrian network and do not allow to communicate directly with local authorities, the business model estimates that there are many benefits of implementing SPN and that far outweigh the costs.

4 Prototype Implementation

SPN was developed, applied and tested considering the central areas of two medium-sized European cities: Porto and Bologna (Fig. 6). Porto is the second largest city in

Portugal and has a population of about 250 thousand inhabitants. Bologna is the capital of the Emilia-Romagna region in Italy and has about 390 thousand inhabitants. Both Porto and Bologna were walled cities with a long and rich history and culture. Based on the outstanding universal value of the urban fabric and its many historic buildings bearing remarkable testimony to the development over the past thousand years, the historic centre of Porto was classified by UNESCO as a World Heritage Site. Bologna is also an old city and contains an immense wealth of significant medieval, renaissance, and baroque artistic monuments. The city is particularly famous by the monuments and by the extensive porticoes and arcades that covered part of the city centre. The conditions provided to pedestrians in both cities are quite different, but both cities are engaged in improving walkability, particularly in the central areas. The prototype implementation of SPN includes SCM cloud engines for Bologna and Porto and their integration with SPA mobile applications. Thus, it is expected that SPN can improve walkability in both cities.

Fig. 6. Street network and central areas of Bologna and Porto.

5 Conclusions

Walking is one of the least expensive and most broadly accessible and sustainable mode of transport. Therefore and, as highlighted by Jabbari et al. [64], the quality of life in cities depends on the existence of suitable conditions to walk. The main goal of this paper was to enhance the concept of walkability in cities developed in the context of the undergoing project SPN - Smart Pedestrian Net. More particularly, the paper explains the concept of Smart Pedestrian Network and the multiple dimensions attached to it to improve walkability. Actions to promote walkability have mainly concentrated on making this sustainable form of travel easy and attractive by improving the general quality of built environments [65]. SPN argues that improving walkability and developing smart pedestrian networks requires a broader and multi-linked integrated approach. The proposed multi-linked approach is complementary supported on three components: urban planning, technological development and business case.

In the planning domain, SPN is supported on a GIS-based multi-criteria model. This model was built to assess the walkable conditions provided to pedestrians and to support the adoption of planning guidelines for improving walkability. The model can be distinguished by the existing multi-criteria assessment methods for the number of criteria involved (23) at both micro and mesoscales, by defining how each criterion should be evaluated and by involving the pedestrians in weighting each criterion. The goal is to make an assessment as much objective as possible and adopt a people-oriented approach: in each city, the model is adjustable according to the pedestrian's perceptions, because criteria with impact on walking are changeable from city to city. Moreover, the urban component is innovative because the criteria more valued were used as routing options in the SPN app. By identifying the streets more and less walkable and the respective causes, the model can be helpful for guiding planning policies to improve walkability and to provide an interconnected (network) set of sidewalks providing good conditions to pedestrians.

To strengthen the smart component of SPN, the project is also supported on the development of a new mobile app (Smart Pedestrian Assistant) to improve the walkability of citizens and tourists. In relation to the existing tools, the SPA is an innovative tool. As emphasised by Papageorgiou et al. [66], many of the existing routing systems were developed for driving or just for monitoring physical activity. However, common concepts used in car navigation tools are inappropriate for pedestrians, mainly because the complexity of human spatiotemporal behaviour requires other route functions than the shortest path. The designed SPA navigation tool has great potential to fill the gap between pedestrian needs and the existing navigation tools due to its useful and innovative features. The app is solely focused on pedestrian use and will enable routing based on the users' preferences and needs. Depending on the pedestrian evaluation (urban planning component), routing options can include, for instance, flat routes, green routes, safe routes, among other possible options. The app will support adjustable functions that can be used according to the characteristics of each user, their preferences in terms of the type of walk, its duration, distance, visits of various points of interests and quick changing of the route to address specific situations. Moreover, the tool is designed to have the ability to communicate directly with the information systems of the city administration to accumulate statistics and to provide information about the pedestrian conditions and urgent or planned events.

Finally, the business component has two goals. Firstly, to disseminate the several benefits (externalities) of walking. The goal is to estimate the benefits of walking instead of driving a car in daily urban trips. Positive externalities include travel time, fuel costs, harmful emissions and health benefits. The dissemination of the benefits of walking is vital for creating a paradigm shift in urban mobility, by promoting a walking mindset and supporting policies and investments to improve the conditions provided to pedestrians. Secondly, a business model was also developed for the SPA app. Preliminary results show that public would be interested in using the app, that could create value for multiple stakeholders namely for users (economic and health benefits), municipalities/public entities (less pollution, healthcare and cleaning costs), and insurance companies (fewer healthcare costs). Promoting the SPA app and involving key stakeholders, particularly those interested in promoting health, increased quality of life, sustainability and urban development is vital for disseminating the tool, attracting funds and increasing market share. In sum, SPN intends to improve walkability by considering three interlinked components that include quantitative and qualitative multiscale assessment, technological development and business components. In the future, the authors expect to present in detail the main findings resulting from the prototype implementation of SPN in Porto and Bologna.

Acknowledgments. FCT co-financing (ENSUF/0004/2016) - The authors gratefully acknowledge ERA-NET Cofund Smart Urban Futures for funding the research project SPN - Smart Pedestrian Net. The authors also acknowledge the national agencies for science, research and technology from Portugal, Italy, Austria and Cyprus for co-funding the project.

https://jpi-urbaneurope.eu/project/smart-pedestrian-net.

References

1. Jacobs, J.: The Death and Life of Great American Cities. Randome House, New York (1961)
2. Hooftman, N., Messagie, M., Mierlo, J., Coosemans, T.: A review of the European passenger car regulations – real driving emissions vs local air quality. Renew. Sustain. Energy Rev. **86**, 1–21 (2018)
3. Nanaki, E., et al.: Environmental assessment of 9 European public bus transportation systems. Sustain. Cities Soc. **28**, 42–52 (2017)
4. Berke, E., Koepsell, T., Moudon, A., Hoskins, R., Larson, E.: Association of the built environment with physical activity and obesity in older persons. Am. J. Public Health **97**(3), 486–492 (2007)
5. UN-United Nations: World Urbanization Prospects, the 2014 Revision Highlights. Department of Economic and Social Affairs, New York (2015)
6. Motoaki, Y., Daziano, R.: A hybrid-choice latent-class model for the analysis of the effects of weather on cycling demand. Transp. Res. Part A **75**, 217–230 (2015)
7. EC - European Commission: White paper: roadmap to Single European Transport Area, Report 144. EC, Brussels (2011)
8. Scarinci, R., Markov, I., Bierlaire, M.: Network design of a transport system based on accelerating moving walkways. Transp. Res. Part C **80**, 310–328 (2017)
9. Sachs, J.: From millennium development goals to sustainable development goals. Lancet **379**, 2206–2211 (2012)
10. Sopjani, L., Stier, J., Ritzén, S., Hesselgren, M., Georén, P.: Involving users and user roles in the transition to sustainable mobility systems: the case of light electric vehicle sharing in Sweden. Transp. Res. Part **71**, 207–221 (2019)
11. Garau, C., Masala, F., Pinna, F.: Cagliari and smart urban mobility: analysis and comparison. Cities **56**, 35–46 (2016)
12. Winters, M., Brauer, M., Setton, E.M., Teschke, K.: Mapping bikeability: a spatial tool to support sustainable travel. Environ. Plan. B Plan. Des. **40**, 865–883 (2013)
13. Pooley, C., et al.: Policies for promoting walking and cycling in England: a view from the street. Transp. Policy **27**, 66–72 (2013)
14. Pucher, J., Buehler, R.: Making cycling irresistible: lessons from The Netherlands, Denmark and Germany. Transp. Rev. **28**(4), 495–528 (2008)

15. Marqués, R., Hernández-Herrador, V., Calvo-Salazar, M., García-Cebrián, J.: How infrastructure can promote cycling in cities: lessons from Seville. Res. Transp. Econ. **53**, 31–44 (2015)
16. Southworth, M.: Designing the walkable city. J. Urban Plan. Dev. **131**(4), 246–257 (2005)
17. Shafray, E., Kim, S.: A study of walkable spaces with natural elements for urban regeneration: a focus on cases in Seoul, South Korea. Sustainability **9**, 587 (2017)
18. Cervero, R., Kockelman, K.: Travel demand and the 3Ds: density, diversity, and design. Transp. Res. Part D Transp. Environ. **2**(3), 199–219 (1997)
19. Cervero, R., Sarmiento, O., Jacoby, E., Gomez, L., Neiman, A.: Influences of built environments on walking and cycling: lessons from Bogotá. Int. J. Sustain. Transp. **3**(4), 203–226 (2009)
20. Frank, L., et al.: The development of a walkability index: application to the neighbourhood quality of life study. Br. J. Sports Med. **44**, 924–933 (2010)
21. Su, S., Pi, J., Xie, H., Cai, Z., Weng, M.: Community deprivation, walkability, and public health: highlighting the social inequalities in land use planning for health promotion. Land Use Policy **67**, 315–326 (2017)
22. Ewing, R., Hajrasouliha, A., Neckerman, K., Purciel-Hill, M., Greene, W.: Streetscape features related to pedestrian activity. J. Plan. Educ. Res. **36**(1), 5–15 (2016)
23. Yin, L.: Street level urban design qualities for walkability: combining 2D and 3D GIS measures. Comput. Environ. Urban Syst. **64**, 288–296 (2017)
24. Galpern, P., Ladle, A., Uribe, F., Sandalack, B., Doyle-Baker, P.: Assessing urban connectivity using volunteered mobile phone GPS locations. Appl. Geogr. **93**, 37–46 (2018)
25. Taleai, M., Amiri, E.: Spatial multi-criteria and multi-scale evaluation of walkability potential at street segment level: a case study of Tehran. Sustain. Cities Soc. **31**, 37–50 (2017)
26. Habibian, M., Hosseinzadeh, A.: Walkability index across trip purposes. Sustain. Cities Soc. **42**, 216–225 (2018)
27. Cain, K., et al.: Developing and validating an abbreviated version of the Microscale Audit for Pedestrian Streetscapes (MAPS-Abbreviated). J. Transp. Health **5**, 84–96 (2017)
28. Aghaabbasi, M., Moeinaddini, M., Shah, M., Asadi-Shekari, Z., Kermani, M.: Evaluating the capability of walkability audit tools for assessing sidewalks. Sustain. Cities Soc. **37**, 475–484 (2018)
29. Shatu, F., Yigitcanlar, T.: Development and validity of a virtual street walkability audit tool for pedestrian route choice analysis—SWATCH. J. Transp. Geogr. **70**, 148–160 (2018)
30. Stewart, O., et al.: Secondary GIS built environment data for health research: guidance for data development. J. Transp. Health **3**, 529–539 (2016)
31. Rigolon, A.: Parks and young people: an environmental justice study of park proximity, acreage, and quality in Denver, Colorado. Landsc. Urban Plan. **165**, 73–83 (2017)
32. Ruiz-Padillo, A., Pasqual, F., Uriarte, A., Cybis, H.: Application of multi-criteria decision analysis methods for assessing walkability: a case study in Porto Alegre, Brazil. Transp. Res. Part D **63**, 855–871 (2018)
33. Nasar, J., Holloman, C., Abdulkarim, D.: Street characteristics to encourage children to walk. Transp. Res. Part A **72**, 62–70 (2015)
34. Sanders, R., Cooper, J.: Do all roadway users want the same things? Transp. Res. Rec. J. Transp. Res. Board **2393**, 155–163 (2013)
35. Hall, M., Ram, Y.: Measuring the relationship between tourism and walkability? Walk Score and English tourist attractions. J. Sustain. Tour. **27**(2), 223–240 (2019)
36. Park, S., Choi, K., Lee, J.: To walk or not to walk: testing the effect of path walkability on transit users' access mode choices to the station. Int. J. Sustain. Transp. **9**(8), 529–541 (2015)
37. Battista, G., Manaugh, K.: Stores and mores: toward socializing walkability. J. Transp. Geogr. **67**, 53–60 (2018)

38. Jabbari, M., Fonseca, F., Ramos, R.: Combining multi-criteria and space syntax analysis to assess a pedestrian network: the case of Oporto. J. Urban Des. **23**(1), 23–41 (2018)
39. Giles-Corti, B., et al.: School site and the potential to walk to school: the impact of street connectivity and traffic exposure in school neighborhoods. Health Place **17**, 545–550 (2011)
40. Blecic, I., Canu, D., Cecchini, A., Congiu, T., Fancello, G.: Walkability and street intersections in rural-urban fringes: a decision aiding evaluation procedure. Sustainability **9**, 883 (2017)
41. Bahrainy, H., Khosravi, H.: The impact of urban design features and qualities on walkability and health in under-construction environments: the case of Hashtgerd New Town in Iran. Cities **31**, 17–28 (2013)
42. Lee, J., Zegras, C., Ben-Joseph, E.: Safely active mobility for urban baby boomers: the role of neighborhood design. Accid. Anal. Prev. **61**, 153–166 (2013)
43. Galanis, A., Eliou, N.: Evaluation of the pedestrian infrastructure using walkability indicators. WSEAS Trans. Environ. Dev. **12**(7), 385–394 (2011)
44. Dhanani, A., Tarkhanyan, L., Vaughan, L.: Estimating pedestrian demand for active transport evaluation and planning. Transp. Res. Part A **103**, 54–69 (2017)
45. Wei, Y., Xiao, W., Wen, M., Wei, R.: Walkability, land use and physical activity. Sustainability **8**, 65 (2016)
46. Conticelli, E., Maimaris, A., Papageorgiou, G., Tondelli, S.: Planning and designing walkable cities: a smart approach. In: Papa, R., Fistola, R., Gargiulo, C. (eds.) Smart Planning: Sustainability and Mobility in the Age of Change. GET, pp. 251–269. Springer, Cham (2018). https://doi.org/10.1007/978-3-319-77682-8_15
47. Ewing, R., Handy, S.: Measuring the unmeasurable: urban design qualities related to walkability. J. Urban Des. **14**(1), 65–84 (2009)
48. Koh, P., Wong, Y.: Comparing pedestrians' needs and behaviours in different land use environments. J. Transp. Geogr. **26**, 43–50 (2013)
49. Guo, Z., Loo, B.: Pedestrian environment and route choice: evidence from New York City and Hong Kong. J. Transp. Geogr. **28**, 124–136 (2013)
50. Bernhoft, I., Carstensen, G.: Preferences and behaviour of pedestrians and cyclists by age and gender. Transp. Res. Part F **11**, 83–95 (2008)
51. Hatamzadeh, Y., Habibian, M., Khodaii, A.: Walking behavior across genders in school trips, a case study of Rasht, Iran. J. Transp. Health **5**, 42–54 (2017)
52. Delmelle, E., Delmelle, E.: Exploring spatio-temporal commuting patterns in a university environment. Transp. Policy **21**, 1–9 (2012)
53. Lois, D., Monzón, A., Hernández, S.: Analysis of satisfaction factors at urban transport interchanges: measuring travellers' attitudes to information, security and waiting. Transp. Policy **67**, 49–56 (2018)
54. Moniruzzaman, M., Farber, S.: What drives sustainable student travel? Mode choice determinants in the Greater Toronto Area. Int. J. Sustain. Transp. **12**(5), 367–379 (2018)
55. Fu, L., Cao, S., Shi, Y., Chen, S., Yang, P., Fang, J.: Walking behavior of pedestrian social groups on stairs: a field study. Saf. Sci. **117**, 447–457 (2019)
56. Oses, U., Rojí, E., Gurrutxaga, I., Larrauri, M.: A multidisciplinary sustainability index to assess transport in urban areas: a case study of Donostia-San Sebastian, Spain. J. Environ. Planning Manage. **60**(11), 1891–1922 (2017)
57. Ribeiro, P., Fonseca, F., Santos, P.: Sustainability assessment of a bus system in a mid-sized municipality. J. Environ. Planning Manage. (2019). https://doi.org/10.1080/09640568.2019. 1577224
58. Lin, J., Wei, Y.: Assessing area-wide bikeability: a grey analytic network process. Transp. Res. Part A **113**, 381–396 (2018)
59. Díaz, R., Muñoz, L., González, D.: The business model evaluation tool for smart cities: application to SmartSantander use cases. Energies **10**, 262 (2017)

60. Papageorgiou, G., Maimaris, A.: Towards the development of Intelligent Pedestrian Mobility Systems (IPMS). In: ICELTICs 2017 International Conference on Electrical Engineering and Informatics, Banda Aceh, Indonesia, 18–20 October (2017)

61. Qiu, L., He, L.: Bike sharing and the economy, the environment, and health-related externalities. Sustainability **10**, 1145 (2018)

62. Papageorgiou, G., Petrakis, C., Ioannou, N., Zagarelou, D.: Effective business planning for sustainable urban development: the case of active mobility. In: 14th European Conference on Innovation and Entrepreneurship, Kalamata, Greece, 19–20 September (2019)

63. Papageorgiou, G., Efstathiadou, T., Efstathiades, A., Maimaris, A.: Promoting active transportation via information and communication technologies. In: 18th IEEE International Conference on Smart Technologies, Novi Sad, Serbia, 1–4 July (2019)

64. Jabbari, M., da Fonseca, F.P., Ramos, R.A.R.: Assessing the pedestrian network conditions in two cities: the cases of Qazvin and Porto. In: Arefian, F.F., Moeini, S.H.I. (eds.) Urban Heritage Along the Silk Roads. TUBS, pp. 229–245. Springer, Cham (2020). https://doi.org/10.1007/978-3-030-22762-3_15

65. Kim, S., Park, S., Lee, S.: Meso or micro-scale? Environmental factors influencing pedestrian satisfaction. Transp. Res. Part D **30**, 10–20 (2014)

66. Papageorgiou, G., Demetriou, D., Balamou, E., Maimaris, A.: Market Research and concept study for a smart pedestrian network application. In: PETRA 2018 Conference, Corfu, Greece, 26–29 June (2018)

Building Future Societies? A Brief Analysis of Braga's School Bus Project

Emília Araújo[1]([⊠]) [iD], Márcia Silva[1] [iD], Rita Ribeiro[1] [iD], and Filipa Corais[2] [iD]

[1] Communication and Society Research Centre, University of Minho, Braga, Portugal
emiliararaujo@gmail.com
[2] Municipality of Braga, Braga, Portugal

Abstract. This paper seeks to demonstrate the importance of implementing transport policies that meet home to school (and vice versa) mobility needs of children and young people. Over the last few years families' mobility has been given increasing attention. One of the most investigated aspects is home to work mobility. Nowadays there is an urgent need to study and propose intervention measures regarding children's mobility to school. In fact, studies show that the car is the preferred mode of transportation in Portugal. This trend has undoubtedly negative social, environmental and health consequences. Based on an evaluation study on the School Bus project in Braga (a research- intervention activity of BUILD-Braga Urban Innovation Laboratory Demonstrator), this paper discusses the difficulties in implementing sustainable mobility initiatives, pointing out anticipatory measures that can be taken in medium-sized cities in order to stop the expansion of children's transportation by car, and give them back spaces in the cities.

Keywords: Urban mobility · Future · City · Children · School bus · BUILD

1 Introduction

This paper aims to show the importance of implementing transport policies for children and young people that meet the needs of families and, at the same time, actively contribute to the much urgent change to more sustainable transport practices, particularly in cities with high population density and where circulation has become progressively chaotic due to the increase in vehicle circulation. It is based on the analysis of the available literature on school transport and the implementation of sustainable mobility for children and young people, and addresses the School Bus project, implemented in the city of Braga, in the North of Portugal.

There are several reasons for thinking about transportation of children and young people to school as a social and sociological problem. Among those reasons, the following ought to be stressed: (i) the regular and growing use of cars and the consequent rush-hour traffic chaos in the cities, which are the cause of accidents, and endless congestions, as well as considerable delays, which expose children and families to stress and anxiety; (ii) the weak provision of public transport and the non- adequacy of these to the mobility needs of families with school-age children and young people, as regards routes

© ICST Institute for Computer Sciences, Social Informatics and Telecommunications Engineering 2020
Published by Springer Nature Switzerland AG 2020. All Rights Reserved
P. Pereira et al. (Eds.): SC4Life 2019, LNICST 318, pp. 143–153, 2020.
https://doi.org/10.1007/978-3-030-45293-3_11

and schedules, among others; (iii) the distance between schools and the families' place of residence and their motivation to add their rides to work with the ones to school and; (iv) the low or total absence of participation of populations, schools, children and young people in the definition of transport policies aimed at quality of life and development of sustainable cities.

In this paper, we find it crucial that the transport of children and young people to and from school becomes a subject of collective and political concern, which involves rethinking and planning urban mobility holistically, not leaving decisions only up to families or schools. Accordingly, we propose the use of transdisciplinary approaches to children and young people's mobility, by suggesting there is a need to help practitioners develop innovation and entrepreneurial actions that are responsible and sustainable, in order to safeguard an alternative and better future for urban areas.

This paper is structured as follows. First, there is a brief introduction to the way social sciences, and particularly sociology, have been working on the phenomenon of children's mobility to and from school, considering that children's transport is a major issue in modern societies. Then, we describe the methodological steps developed during a research and evaluation project on the implementation of the school bus in the city of Braga. Afterwards we describe how families and children joined the school bus initiative and analyze its main positive and negative points. Finally, we discuss some practical considerations about the way children's mobility can be managed in order to reduce existing traffic congestion due to the massive use of cars in the city.

2   Urban Mobility: From Difficulties to Opportunities

Society's car dependency is difficult to revert [1–3] and entails several implications for children's lives [4]. In fact, children miss out on the opportunity to gain skills and autonomy, as well as the opportunity for social interactions, when circulating almost exclusively in cars. Furthermore, families waste time that could be used in more stimulating and fruitful activities. Besides health problems related to high levels of air pollution, and other diseases brought about by the increasing accelerated pace of life, cars are also related to the abovementioned increase in traffic congestion. It is a fact that the problem of traffic in cities, particularly in southern European countries, has been addressed by putting a great emphasis on the need to transition to electric and autonomous cars. However, this transition does not solve any major effects caused by cars. Instead, it only reshapes urban spaces' current challenges.

Actually, it is urgent to integrate school mobility and mobility of families in urban policies, by promoting means of movement and transport which, at the same time, empower children and allow them to enjoy public space and ensure a safe circulation between school, home and other places (for example, extracurricular activities) in a more sustainable way. This can be achieved in several ways, including by implementing the use of public transport, or by encouraging safe walking routes for children in their way to school. In any case, families and children needs must always be met. Most importantly, giving them back public space and teaching them how to reuse the time and space of transportation and travel, is a key factor for quality of life in the cities of the future.

2.1 The Problem of Mobility in Braga

From the citizen's point of view, assessed through field observation, the circulation in Braga has been getting increasingly more difficult for these last couple of years. People complain about the fact that there is traffic everywhere, circulation is very slow and parking is scarce and expensive.

Braga has currently about 181,000 inhabitants, with 57.8% of the population aged between 25–64, while the population under 25 is about 26%. The city centre is where most of the services, such as schools, health care services, public services, sports centres, among others, are. Braga attracts many people every day (about 48.020 movements) mostly from neighbouring counties: Vila Verde (11.502), Guimarães (10.638) and Barcelos (9.415) [5]. About 27.285 people move from Braga every day. The use of cars in Braga has been growing since 2010 (+17.5%). This is a trend that follows what is happening all over the country [6] (Fig. 1).

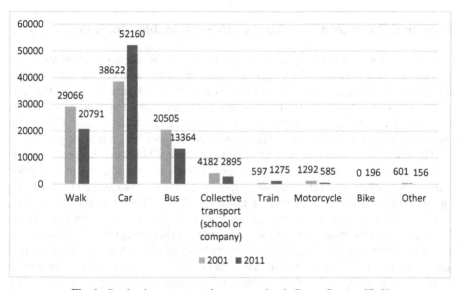

Fig. 1. Predominant means of transportation in Braga Source: [7, 8]

According to Teles [5], cars are used in about 67.3% of the movements within the county, and 79.8% outside the county. The average time spent in commuting home to work or schools is about 15 min (58.6%), with 30.5% of commuters spending 16 to 30 min. Rush hours occur between 8 and 9 am, and between 6 and 7 pm. There are about 499 cars per 1 000 inhabitants in Braga – a value equivalent to the national average. According to the same study [5], and because there are numerous schools in the city centre, car transportation to schools between 2001 and 2011 has grown more than 33%. These numbers clearly confirm the massive use of cars in day-to-day journeys, including home-school movement, and corroborate citizens' perception about mobility in Braga being "chaotic", "slow", "difficult" and "dangerous" (Fig. 2).

Fig. 2. Children crossing a street near school, in Braga (a car prevents them from crossing in the right place)

3 School Buses as a Key Answer to Sustainable Mobility

Nowadays, school buses are seen as an innovative, sustainable alternative in urban mobility. They are used worldwide for transporting students to and from school. Some of their most relevant effects relate to an increase in air quality by reducing CO_2 and other pollutant emissions, a decrease in traffic flow to schools, an increase in children's safety in their way to school and time families spend together. Besides being in tune with current environmental concerns, the fact is that the School Bus goes back a long way.

The first School Bus dates back to 1827, when George Shillibeer, inspired by buses in Paris, took the initiative to London. Despite its Parisian origin, it was in the United States of America that the school transport spread when Wayne Works built horse- drawn carriages - *school hacks* or *kid hacks* - in Indiana. Before having this means of transport, children went to school on foot, by wagon or sled. As car industry developed, Wayne Works took the opportunity to motorize these carriages, shortening travel time to schools. In 1936, the company developed a set of school buses and introduced the first design of school buses in steel. They are traditionally painted in yellow because this color is more visible in the morning and at night. Although the project was not very well accepted by parents, due to some safety issues, it was eventually considered one of the safest modes of school transport in the USA [9–11].

More recently, some cities have implemented school bus services, both on a permanent basis or as pilot initiatives. Some of the projects carried out mainly in Europe are mentioned below. At the Deutsche Schule, in London, the School Bus service has four routes. This service is available in the morning and afternoon. Costs associated with the service are incurred by the parents [12]. According to the newspaper *The Guardian* [13], some schools in London, Manchester and Edinburgh are taking steps to prohibit automobile traffic nearby schools. The Scottish government is committed to encouraging active transportation to and from school and reducing car dependence. In Scotland, during the "rush hour" - in the morning and in the afternoon- it is estimated that one in five cars on the road are busy transporting children to school. The government is interested in reducing the use of cars as a way to improve the health and well-being of children and young people, while reducing congestion and decreasing CO2 emissions. In 2009, the city of Lisbon carried out a mobility project aimed at implementing a bus system for public schools. This project aimed to reduce student mobility costs. The municipality traced some specific circuits (home-school and school-home), for students living more than 6 min' walk from school. Routes were established so that children would not have to walk long to get to the bus. The buses "Alfacinhas" had the capacity for 24 children, with average travel time of 15 min. Each bus had someone who was responsible for picking up and delivering each child to their parents, at the previously designated 'bus stop' [14]. In Canada, the Red Deer Catholic Regional Schools have implemented the School Bus service. Service costs are incurred by the families. Regarded as an extension of the school facilities, students cannot eat, drink, or stand when inside the buses. This service includes a mobile application that allows parents to know if the school bus is delayed or if, for some reason, it has been cancelled [15]

In general, studies about school buses show that they can be an effective alternative to using the car, or to walking to school. Besides the positive effects in terms of families' spent time and money, air quality and urban traffic, there are also some advantages associated to the well-being of children, such as: (1) the strengthening of personal contact, thus favoring sociability, as children have the possibility of interacting with colleagues of different ages and conditions, fostering feelings of trust and self-esteem; (2) the development of children's autonomy; (3) the safety of children, as public transportation is safer than cars, and school buses have low accident rates, most of which do not imply lethal injuries [16–20]. The more children are on buses, the fewer cars are on the roads.

4 The School Bus Project in Braga

The School Bus Project in Braga aims at reducing traffic both in the city centre and in the main entrances of the city, especially during morning rush hour. It has been designed and implemented by the Municipality of Braga in a partnership with a wider project – Braga Urban Innovation Laboratory Demonstrator[1] (BUILD). The project encompasses six schools: Colégio Leonardo Da Vinci, Colégio D. Diogo de Sousa, Colégio Teresiano, EB 2.3 Francisco Sanches, EB 2.3 André Soares and Conservatório de Música Calouste Gulbenkian. After the first experience in 2017, the school bus has been steadily operating since September 2018, with 413 children enrolled. Parents leave their children at one of the four interfaces (Municipal Stadium, Maximinos, Minho Center, Tenões) and they are, then, transported by bus to the schools' facilities. Being an important measure of the urban mobility policy, the Municipality undertakes all the costs of this program, so all families may benefit from free transportation, regardless of social and economic status (Fig. 3).

---

[1] Braga Urban Innovation Laboratory Demonstrator (BUILD) is a project of the Municipality of Braga with the following scientific partners: University of Minho, International Iberian Nanotechnology Laboratory and Centre for Computer Graphics; funded by the Fundo Ambiental - Ministry of the Environment.

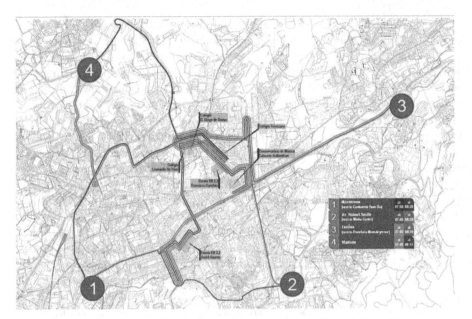

Fig. 3. School bus interfaces and routes in Braga Source: Municipality of Braga

5   Methodological Note

This paper is based on bibliographic research on the use and history of the school bus and the results of a study conducted in the city of Braga on the School Bus project. The methodological choices allowed us to assess the project, as well as to analyse families' mobility practices and major concerns presented by some important stakeholders around thinking and reframing school mobility. A mixed method was followed, using qualitative and quantitative research techniques. Methodology encompassed systematic observation, focus groups, interviews and a survey.

The interviews involved some important players including school principals, parents and representatives of other entities, such as the police, the parish and the county council. The semi structured interviews were done according to a guide integrating the agents' concerns, personal experience and the identification of measures they suggested to solve problems related to school mobility. The interviews were recorded. After the transcription they were analysed according to content analysis procedures using MaxQda software.

The focus groups were conducted with children. Part of them were school bus users, others were not. These sessions with the children lasted for a maximum of 90 min. Children were asked to draw their main mean of transportation and then challenged to speak about it and share some ideas about the type of things they would change in their daily habits linked to mobility. Following data collection and systematization in a database, content analysis was used to extract information about three dimensions: the most frequent means of transportation to and from school, desire and intention of changing something in the trip.

The survey was applied to parents of children registered to use the School Bus. The questionnaire consisted of 27 questions, about sociodemographic information, safety perception on the areas surrounding schools, school mobility practices and satisfaction with the use of School Bus. The questionnaires were handed out to children in order to be filled in by their parents.

96 (out of the 130 delivered surveys) responses were received, involving 127 children, which represents a high response rate, ensuring considerable reliability of the conclusions. SPSS software was used to analyse the data.

6 Project Evaluation

6.1 School Directors

Mostly, school directors highlight the fact that the project contributes to creating new mobility habits, because "children who use public transport now will also use it later in life" (director of Colégio Leonardo da Vinci). In the interviews, directors stressed three main positive achievements: (1) student's punctuality, as children who use the School Bus do not arrive late to school. According to the director of Colégio D. Diogo de Sousa, the initiative "disciplines and helps students stick to schedules"; (2) pedagogical advantage for participating children as they arrive to school more willing and motivated for pedagogical activities than children who come by car. Children who take the school bus "arrive in a happier mood, more motivated to take classes. Children who arrive late with their parents, after working through the rush hour, directly into the classroom do not have the same healthy disposition" (Director of Colégio Leonardo da Vinci); (3) a significant decrease of traffic in the area nearby school, making it a more pleasant place for users (security, air quality, and calmness). However, directors also mention some negative aspects, namely the fact that the School Bus is not available for children of all ages, families are not yet aware of its advantages and struggle with the change in behaviour regarding the use and parking of cars near schools.

6.2 Parents

Even if they evaluate the School Bus very positively (82% rate it as excellent or very good and 98% intend to continue using the service in the future), parents seem to have some mixed feelings concerning the need to leave the children at the bus stop or taking them by car. According to the data gathered in the parents' evaluation survey, these contradictions are due to subjective representations and evaluation about children's safety, as well as their instinct to protect kids and a wish to spend as much time as they can with them, even if in traffic jams. The evaluation survey results show that parents are highly satisfied with the project as they feel children are more "independent, autonomous, and safe", they meet (new) friends, "they like to go with their mates, it is comfortable and makes them feel free and they are able to arrive earlier to school", which is almost impossible when traveling by car.

Parents also say that the School Bus helps them saving time in traffic as they are able to avoid the usual jams near schools and go directly to the workplace. Punctuality and quality of life are also mentioned. The table below sums up the most positive aspects mentioned by the parents (Table 1).

Table 1. Positive and negative aspects mentioned by parents in their evaluation of the school bus project in Braga

Positive aspects	Negative aspects
• Decrease in traffic and pollution	• The school bus should also ensure the bus ride home
• Strengthening of friendship ties with neighbors, both for children and their families	• Schedules should be more adequate, so that children should not have to get up so early in the morning
• Punctuality	• More routes and schedules should be available, for more flexibility
• Time saving for the family	

Overall, parents are very critical of the efficiency, availability and organization of public and school transportation in the city of Braga, not to mention the need for a service tailored to their daily routines. Besides, parents strongly claim that the spaces surrounding schools should be more secure and car-free, and users (especially parents) should avoid inappropriate behaviour such as parking in the middle of the street, or outside the parking spaces.

6.3   Children

Despite being very young (6 to 15), most of the School Bus users have strong opinions about this type of home-to-school transport. Because of their age, qualitative data gathering was an obvious choice. Focus groups took place in schools and sessions encompassed debating, writing and drawing about mobility. Participants were both School Bus users and non-users. Interestingly, 60% of the latter said they would like to join it. For those already using this service, positive opinions are over 95% and the main reasons for that are the urge to protect the environment, socializing with friends and arriving to classes on time (Fig. 4).

7   Children's Right to the City: What Still Needs to be Done

Somewhat unexpectedly, and contrary to the current environmental challenges and recent problems in urban areas, families seem to be increasingly dependent on cars. The pace of life, the need to respond to various demands located in different spaces, the distances between home and the workplace, the inefficiency of the public transportation network, the ever-increasing supply of cars are some of the reasons used to justify the increase in car users [21]. One of the factors that contribute greatly to the use of cars is also

Fig. 4. Child's drawing depicting the school bus ("Less mobile phones, more friends, less cars")

the number of after school activities children are involved in every day, which implies traveling both from home and from school. These activities include sports, music classes, foreign languages classes, among others [21, 22].

The analysis of the existing literature and the qualitative research conducted in the context of the School Bus project in Braga shows that the transportation of children and young people is a major issue in "smart cities" agenda, not only because of decarbonisation and quality of life in city centres, but also because short daily home- school-work commuting is deeply connected to societal challenges such as territorial and social cohesion, citizenship and participation, and bottom-up strategies for urban ecosystems [23–25].

To this end, measures that bring children and young people to the centre of the analysis contribute to thinking about the public space from the perspective and experience of those who will shape the future of our cities [26, 27]. For this purpose, social research ought to focus on the appropriation of spaces and commuting patterns of different social groups. Additionally, attention should be drawn to the risks and dangers associated with urban

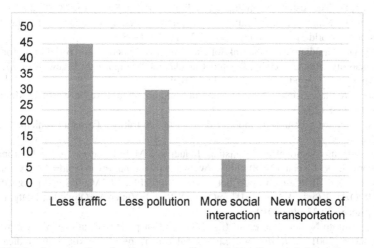

Fig. 5. Children's suggestions for change regarding mobility Source: Information provided by children's focus groups

circulation, intervention measures that favour the participation of children in the design of the city (see Fig. 5) and the routes to be carried out daily.

8 Concluding Remarks

In this text, we intended to present the main advantages of using the school bus on the routes to school, considering the importance of including children's mobility as a priority in the planning and design of cities in the future.

Transporting children to schools is a matter of great relevance in the analysis of mobility patterns of families [28] and has a significant impact not only on families' daily lives, but also on the organization of the territory because home to school commuting represents a weighty part of traffic congestion in the cities, especially during rush hours, both in the morning and by the end of the day [29].

However, using the car has a number of consequences, including for the health and well-being of children. Some are visible (for example, public spaces overcrowded with cars and the insecurity it entails). Others are more implicit, such as hindrances to children's autonomy or lack of physical activity.

Given the School Bus project's results, we may conclude that the objectives have been partially achieved, by having demonstrated the relevance of children and families' support for social and environmental sustainability regarding circulation in urban areas.

Funding. Funded by FCT - Fundação para a Ciência e Tecnologia, within the UID/CCI/ 00736/2019 project scope. Funded by Fundo Ambiental, project BUILD-Braga Urban Innovation Laboratory Demonstrator.

References

1. Banister, D.: The sustainable mobility paradigma. Transp. Policy **15**, 73–80 (2008)
2. Meaton, J., Kingham, S.: Children's perceptions of transport modes: car culture in the classroom? World Transp. Policy Pract. **4**(2), 12–16 (1998)
3. O'Brien C.: Children: a critical link for changing driving behaviour. York Centre for applied sustainability. York University, Toronto, Canada (2000). http://www.bikewalk.org/pdfs/child rendrivingbehavior.pdf. Accessed 21 June 2019
4. Mackett, R.: Increasing car dependency of children: should we be worried? Munic. Eng. **151**(1), 29–38 (2002)
5. Teles, P. (coord.): Estudo de mobilidade e gestão de tráfego. Câmara Municipal de Braga (2018)
6. INE - Instituto Nacional de Estatística: Inquérito à Mobilidade nas Áreas Metropolitanas do Porto e de Lisboa (2018). https://www.ine.pt/xportal/xmain?xpid=INE&xpgid=ine_p ublicacoes&PUBLICACOESpub_boui=349495406&PUBLICACOESmodo=2&&fbclid= IwAR2QzUZK0mUSEdKySZe1HqmObblKWR62vIyVhtVAAxrQhyNllna-DDfp2bk&xlan g=pt. Accessed 25 June 2019
7. INE - Instituto Nacional de Estatística, Meio de transporte mais utilizado nos movimentos pendulares (2001). https://www.ine.pt/xportal/xmain?xpid=INE&xpgid=ine_indicadores&i ndOcorrCod=0000684&contexto=bd&selTab=tab2. Accessed 25 June 2019

8. INE - Instituto Nacional de Estatística, Meio de transporte mais utilizado nos movimentos pendulares (2011). https://www.ine.pt/xportal/xmain?xpid=INE&xpgid=ine_indicadores&indOcorrCod=0007093&contexto=bd&selTab=tab2. Accessed 25 June 2019
9. The Newswhell. History of the School Bus. (Blog post). https://thenewswheel.com/history-of-the-school-bus/. Accessed 11 June 2019
10. Transportation and Scholl Busing. The School Bus, History of Pupil Transportation, Issues in Pupil Transportation: https://education.stateuniversity.com/pages/2512/Transportation-School-Busing.html. Accessed 11 June 2019
11. Wagons, Hacks, and Sledges: History of the School Bus. https://www.edgarsnyder.com/blog/2015/08/history-of-school-bus.html. Accessed 11 June 2019
12. Deutsche Shule London, School Bus. https://www.dslondon.org.uk/admissions/school-bus. Accessed 08 June 2019
13. The Guardian, UK schools banning school run to protect pupils from air pollution. https://www.theguardian.com/environment/2018/jul/13/uk-schools-move-to-ban-the-school-run-to-protect-pupils-from-air-pollution. Accessed 08 June 2019
14. Martíneza, L., Viegasa, J.: Design and deployment of an innovative school bus service in Lisboa. Procedia Soc. Behav. Sci. **20**, 120–130 (2011)
15. Red Deer Catholic Regional Schools, School Bus Information. https://www.rdcrs.ca/schools/school-bus-information. Accessed 08 June 2019
16. Tamanaha, S.: Design para sistema de transporte de estudantes. Dissertação de final de curso, Universidade de São Paulo, Brasil (2014). http://www.fau.usp.br/fauforma/2015/assets/sergio_tamanaha.pdf. Accessed 18 May 2019
17. Yang, J., Peek-Asa, C., Cheng, G., Heiden, E., Falb, S., Ramirez, M.: Incidence and characteristics of school bus crashes and injuries. Accid. Anal. Prev. **41**, 336–341 (2009)
18. Desai, M., Takkalaki, P., Bhapri, M., Marshanalli, A., Malage, G.: Students tracking system for school bus. Int. Res. J. Eng. Technol. **4**(6), 2554–2557 (2017)
19. Kotoulaa, K., Botzoris, G., Morfoulaki, M., Aifandopoulou, G.: The existing school transportation framework in Greece – barriers and problems comparing to other European countries. Transp. Res. **24**, 385–392 (2017)
20. Tenniz, Z., Yilmaz, S.: Attention! school bus. Eur. J. Educ. Stud. **4**(2), 276–288 (2018)
21. Dowling, R.: Cultures of mothering and car use in suburban Sydney: a preliminary investigation. Geoforum **31**, 345–353 (2000)
22. Hjorthol, R., Fyhri, A.: Do organized leisure activities for children encourage car-use? Transp. Res. **43**, 209–218 (2009)
23. Moodie, M., Haby, M., Galvin, L., Swinburn, B., Carter, R.: Cost-effectiveness of active transport for primary school children - walking school bus program. Int. J. Behav. Nutr. Phys. Act. **6**, 1–11 (2009). https://doi.org/10.1186/1479-5868-6-63
24. Nasrudin, N., Nor, A.: Travelling to school: transportation selection by parents and awareness towards sustainable transportation. Procedia Environ. Sci. **17**, 392–400 (2013)
25. Lopes, F., Cordovil, R., Neto, C.: Children's independent mobility in Portugal: effects of urbanization degree and motorized modes of travel. J. Transp. Geogr. **41**, 210–219 (2014)
26. Pojani, D., Stead, D.: Sustainable urban transport in the developing world: beyond megacities. Sustainability **7**, 7784–7805 (2015)
27. Baslington, H.: Travel socialization: a social theory of travel mode behavior. Int. J. Sustain. Transp. **2**, 91–114 (2008)
28. Oliveira, C.S., Cruz, M.: Urban mobility patterns and the use of public transports in Portugal. In: Proceedings Clima 2008. Universidade de Aveiro (2008)
29. Ristell, J., Enoch, M., Quddus, M., Hardy, P.: Expert perspectives on the role of the bus in school travel. Munic. Eng. **166**(1), 53–58 (2013)

City Rankings and the Citizens: Exposing Representational and Participatory Gaps

Ana Duarte Melo^(✉) ⓘ

Communication and Society Research Centre, University of Minho, Braga, Portugal
anamelo@ics.uminho.pt

Abstract. A critical reflection on the purposes, role and performance of city rankings through an holistic communicational approach is at the core of this article. Grounded on a conceptual framework that highlights the contemporary idea of the city—beyond the smart city and more as a co-intelligent, collaborative and co-creative entity, and on the performance outputs of city rankings as territorial and strategic communication tools that actually represent the state of cities, we address the citizens' presence or contribute—as main city stakeholders—to city rankings. In order to make research tangible with a practical component, an exploratory comparative content analysis of three recognized city rankings: the CBI – City Brands Index 2017, the GCR – 2018 Global Cities Report, and the Global Liveability Index 2018—was carried out.

Conclusive notes argue that in order to effectively represent cities, as they are lived, thought and built by their citizens in their everyday, city rankings must rely in more real-time, updated, people's perception centred data, and embed more citizen participation and insights. Moreover, methodology transparency and accountability should be promoted in order to add trust value to city rankings.

Keywords: City rankings · Communication · Citizenship · Participation

1 Introduction

A communicational approach to territories is in order when we address the process of building "smart and liveable cities". The main insight of this article intends to reflect on how territories, particularly cities, engage in communication and on how communication—namely through city rankings, contributes (or not) to better cities. We will focus our attention in particular on city rankings as tools for gauging what is relevant in a city, but also as central communication tools in building the city's image, reputation and representation, that is, in adding symbolic and narrative value to the city. Our perspective draws on the idea of the city made by and for the citizens, a city as the communal capital and knowledge, a smart city that serves, includes and drives the collective. Therefore, the main question we intend to address could be summarized in a straightforward interrogation: do city rankings take the citizens into account through participatory practices?

© ICST Institute for Computer Sciences, Social Informatics and Telecommunications Engineering 2020
Published by Springer Nature Switzerland AG 2020. All Rights Reserved
P. Pereira et al. (Eds.): SC4Life 2019, LNICST 318, pp. 154–169, 2020.
https://doi.org/10.1007/978-3-030-45293-3_12

To assess this, we will analyse a number of city rankings and map some of its constitutive criteria that allegedly include citizens' inputs, whether through direct or indirect participatory practices. Furthermore, we will reflect critically on the use of city rankings as territorial communication management tools, namely on their strategic and managerial value. To set a framework to this research a diversity of notions is summoned, namely, the ideas of city, citizens and participation as constitutive parts of the city, the concepts of smart city, co-intelligent territory, territorial communication management, city brands, image and reputation. We consider them nuclear to understand the subject of city rankings.

What is a city and what does it mean? And how do city rankings contribute to construct that meaning and add value to the city?

2   A Conceptual Framework

2.1   From the *Smart City* to the *Co-intelligent Territory*

The smart city concept emerged in the 1990s, associated with the application of information and communication technologies to cities, therefore embedded within an intelligence framework, whether technology, human, institutional, energy, or data management centred just to name a few. Two main approaches can be found related to smart cities: (1) a management approach, focused on efficiency, resources and flows, crossing a number of areas: energy, mobility, transport, consumption, waste disposal, environmental impact, etc.; and (2) a social innovation approach, in which smart cities improve the lives of citizens and visitors: a smart city collects data from citizens, devices and assets to interact directly with both the community and the city infrastructure; optimize city operations and performance, services and interaction with citizens.

"There is a range of definitions of a smart city, but the consensus is that smart cities utilize IoT sensors, actuators, and technology to connect components across the city. This connects every layer of a city, from the air to the street to underground. It's when you can derive data from everything that is connected and utilize it to improve the lives of citizens and improve communication between citizens and the government that a city becomes a smart city" (Maddox 2018).

As recent as it is, the smart city is a work-in-progress concept, prone to discussion and criticism (Cardullo and Kitchin 2019; Hollands 2008; Rizzon et al. 2017). Hollands (2008) proposes a "preliminary critical polemic" in an effort to clear "[smart cities] definitional impreciseness, numerous unspoken assumptions and a rather self-congratulatory tendency (what city does not want to be smart or intelligent?)" (Hollands 2008, p. 304). The smart city concept involves both an operational, managerial insight and a philosophical and holistic approach to the city (Vanolo 2014).

Several designations emerged that point towards a general definition of what a smart city is: "Intelligent (…) Digital (…) Sustainable (…) Technocity (…) Well-being city" (Dameri 2013, p. 2546); or, in a more dynamic perspective, integrating the notion of change and transformation: "In today's modern urban context, we appear to be constantly bombarded with a wide range of new city discourses like smart, intelligent, innovative, wired, digital, creative, and cultural, which often link together technological informational transformations with economic, political and socio-cultural change" (Hollands 2008, p. 305).

According to Marsal-Llacuna, Colomer-Llinàs and Meléndez-Frigola, "Smart Cities have evolved out of livable, creative, digital and knowledge cities, drawing heavily on the concept of the Sustainable City and having in common a large technological component" (Marsal-Llacuna et al. 2015, pp. 617–618).

Smart cities' critique frequently deals with opposing perspectives, namely techno centred *vs* citizen centred or the private interest *vs* public interest; or long term effects and objectives, namely sustainability (Martin et al. 2018); ethical issues such as privacy (Zhang et al. 2019), big data; freedom (Vanolo 2014) or participation (Cardullo and Kitchin 2019). Many of these critical perspectives converge on an essential question: what does it mean to be a citizen in a city, i.e. what does it mean to be a *smart citizen* in a *smart city*?

Maddox (2018) highlights smart cities' progress and perspectives, "Smart cities are more than a trend—they're the wave of the future because the world is becoming more urban, with 60% of the population expected to live in cities by 2050" (Maddox 2018) and proposes a comprehensive guide to what a smart city is, concluding with a provocative statement: "A truly smart city improves the quality of life for citizens and visitors, and while a smart city can be many things, just as with humans, some are smarter than others" (Maddox 2018).

In a comprehensive proposal Dameri (2013) establishes structural elements, namely, territory, technology, citizens and government as components, and environmental sustainability, quality of life and well being, participation, knowledge and intellectual capital, as objectives of the smart city concept.

Chourabi et al. (2012) identified eight critical factors of smart city initiatives: management and organization, technology, governance, policy context, people and communities, economy, built infrastructure, and natural environment.

In a previous research we proposed a broader insight (Melo 2019), inspired on Tom Atlee's concept of co-intelligence (Atlee 2012, 2014, 2017), to approach territorial "smartness", one based on systematic communicative engagement within the city's stakeholders: "There is no smart territory without smart communication. There is no smart territory without smart citizens who share the vision, essence, projects and planning of territories." (Melo 2019, p. 247). In this article we will focus further on the participation area of Dameri's and on Chourabi's fifth factor—people and communities—as constitutive structural elements of the smart city. This perspective coincides with the conclusions of Hollands (2008) that, to a certain extend, expose the smart city as an impediment in itself:

"In essence the smart progressive city needs and requires the input and contribution of these various groups of people, and cannot simply be labelled as smart by adopting a sophisticated information technology infra-structure or through creating self-promotional websites. Cities are more than just wires and cables, smart offices, trendy bars and luxury hotels, and the vast number of people who live in cities deserve more than just these things. Because the smart city label can work to ideologically mask the nature of some of the underlying changes in cities, it may be a partial impediment toward progressive urban change. Real smart cities will actually have to take much greater risks with technology, devolve power, tackle inequalities and redefine what they mean by smart itself, if they want to retain such a lofty title" (Hollands 2008, p. 316).

In fact, collective intelligence is not a new concept. It has been studied more consistently after the 1980's. Stremtan (2008) wrapped up a definition based on diverse theoretical perspectives: "Collective intelligence is intelligence that emerges from the collaboration and competition of many individuals (...) appears in a wide variety of forms of consensus decision making (...) is the intelligence of connections, of relations (...) These connections induce co-operations, which constitute materialization of collective intelligence (Stremtan 2008, p. 10)".

Furthermore, the author refers to collective intelligence as "the capacity of an organization, of a collective to ask questions and to seek the answers together" (Stremtan 2008, p. 10). Concurrently, Tom Atlee's proposes a holistic, inclusive perspective of what co-intelligence is: "intelligence that takes wholeness, interconnectedness, co-creativity and participation seriously. Co-intelligence is collective, collaborative, synergistic, wise, empathic, heartful, and connected to greater sources of intelligence. It is often marked by how creatively it uses dissonance and diversity." (Atlee 2017, pp. 80–81). Therefore, a co-intelligent city would be grounded on democratic participatory practices, oriented by public wisdom and citizen-led politics and driven by social activism.

These visions, based in interconnectedness and participation, seem crucial to frame both the managerial and the communicative approaches—inspired in participatory and community communication—to territories in the contemporary.

Moreover, discussions on the concept of the city and the right to the city point towards a similar direction: a smart city is a communicative city. And is mostly defined by its citizens.

"The terms global city, smart city, connected city, sustainable city, inclusive city, and green city all encapsulate the collective value placed on a location where many people live and work. All these terms are both descriptive and normative of a reality based upon a gathering of people located in close proximity. A city consists of structures and services, but a city without people is simply an inanimate shell—a dead space. People and their needs define a city" (Drucker and Gumpert 2016, p. 1366).

2.2 Territorial Communication and City Rankings

Braga has currently about 181,000 inhabitants, with 57.8% of the population aged Territorial communication is intrinsic to the city's definition. Through communication the city as to think, project and question it-self on what it wants to communicate, how it wants to be perceived, why and what it wants to communicate. This is essentially strategic thinking, transversal throughout the communication interaction in the construction of the city's brand, image, reputation and symbolic value (Kavaratzis 2019; Kavaratzis and Kalandides 2015; Melo 2018, 2019; Papp-Váry 2011), hence, in the construction of its own identity (Anholt 2007; Ingenhoff et al. 2019).

"Besides the material capabilities of cities, how well they are able to brand themselves plays a large part in their success. The number of tourists, investors, new inhabitants, or the products of the city depend on the success of this self-branding. It also affects the locals and their sense of pride and contentment living in the city. Fortunately for cities, more and more brand models and rankings are available to assess their standing among others in respect to image and to reveal their comparative shortcomings" (Papp-Váry 2011, p. 555).

Territory provides a rich field for research, summoning a diversity of perspectives, including geography, public diplomacy, business and management, social psychology, sociology, urbanism and planning, political communication and communication science, demanding a transdisciplinary approach (Buhmann et al. 2019).

Territorial communication models have been developed and refined to serve the strategic communication of places (Anholt 2007; Buhmann et al. 2019; Kavaratzis 2019; Papp-Váry 2011). The most popular and recognized territorial communication model is certainly Simon Anholt's hexagon. This seminal model created to frame nation-brand communication was later adapted to cities. Also known as de "P's model" is based on five categories: Presence, Place, Pre-requisites, People, Pulse and Potential (Kirchener 2018, p. 2). Such models serve as inspiration for the setting of territorial indexes although open to systematic improvement, as "the challenges of combining the appeal of a holistic approach (…) with the necessity of specific measurement tools have not as yet been solved" (Merkelsen and Rasmussen 2019, p. 81).

Territorial rankings have been established—and used—for a number of reasons, assessment and competiveness being the most obvious ones (Drucker and Gumpert 2016; Merkelsen and Rasmussen 2019; Newburry and Song 2019). They are part of the strategic ammunition territorial managers, communication experts, politicians, urban planners and decision makers in general use to promote, enhance and add value to their nations, regions and cities. They use it to assess territories attractiveness—both theirs and their competitors, thus promoting territorial competitiveness (Chourabi et al. 2012; Giffinger et al. 2010), motivating and rewarding their teams performance, and establishing tangible strategic objectives.

Referring to nation branding attraction conditions, Merkelsen and Rasmussen argue that indexes "serve as catalyst for creating these conditions" (Merkelsen and Rasmussen 2019, p. 77) and they specify three: "a reputational logic (…) a state of competition (…) a simple solution to a complex challenge" (Merkelsen and Rasmussen 2019, pp. 77–79). They conclude by noticing the paradox that what "renders the nation brands index inappropriate as a tool for governments is the exact same quality that makes these rankings attractive to them" (Merkelsen and Rasmussen 2019, p. 81). These paradoxical characteristics seem to fit also the case of city indexes.

There are all types of rankings highlighting economic features, assessing developing metrics, cultural drivers, wellbeing, happiness, participation, transparency, governance and sustainability (Cabannes 2019; McArthur and Robin 2019; Meijering et al. 2014; Okulicz-Kozaryn 2013; Yigitcanlar and Inkinen 2019).

"The proliferation of criteria has led to the propagation of rankings and award programs allowing cities to proudly advertise their recognition as: livable, clean, large, sustainable, green, healthy, and slow. Awards recognize projects, vision, design, and research" (Drucker and Gumpert 2016, p. 1369).

Giving a meaning to data seems to be one of the outcomes and objectives of territorial rankings. Although some criteria might be fairly clear and objective such as the existence of a public transport network or public health care, other criteria such as the level of censorship or corruption might be more difficult to grasp. If aggregating data might be useful to understand some metrics, disaggregating them is frequently the fastest track to understand how narratives around territories are directed, i.e., if a city ranks good in a public health network assessment, does not mean that there is actual access from the population to that resource. Thus, when assessing a ranking, is important to mine the sources, the methods and interpretations offered to that data.

Marsal-Llacuna, Colomer-Llinàs and Meléndez-Frigola focused their work on establishing a set of criteria assembled in "a synthetic indicator to summarize a city's "intelligence" or "smartness"" (Marsal-Llacuna et al. 2015, p. 612). They draw attention towards the need to have physical and non-physical urban data, as much real time as possible, and highlight the need to produce comparable indicators with other rankings. More specifically, the authors insist on the "the need for an "intelligent index" to monitor the Smart Cities initiative" (Marsal-Llacuna et al. 2015, p. 620).

Rankings themselves are object of rating. The Branding Institute CMR, set up a methodology to evaluate rankings based on four criteria: "Relevance & Impact of Ranking; Added Value & Insights of Ranking; Trustworthiness & Intention of Ranking provider and media outlet and Methodology of Ranking" (Casanova et al. 2018).

Nevertheless, being commonly accepted as communication tools central to reputation and image management, territorial rankings have been criticized for being prone to instrumentalization and manipulation. Indexes are "more rhetorical than scientific as they primarily serve purposes of legitimization and persuasion" (Merkelsen and Rasmussen 2019, p. 80). Critique draws mostly on the fact that rankings are both being data driven vs people driven, elitist vs communal—"City rankings are a window onto the projected tastes of a highly specific elite" (O' Sullivan 2019), promoting mainly private interests instead of territorial ones, thus not representing the real life of the city: "it seems mistaken to present all cities as simple clusters of amenities, not when access to them is being subtly discouraged by broader social trends or actively denied to certain groups" (O'Sullivan 2019).

With a provocative tittle—"The Top 6 Reasons to Be Wary of City Rankings, Ranked"—Aarian Marshall draws attention to the controversial and unavoidable issues surrounding the topic of territorial rankings: "Different rankings measure totally different things. (…) Not all cities were included. (…) Some rankings measure reputation—not reality. (…) Some data is hard to find … (…) … While some data is just silly. (…) They're probably trying to sell you something" (Marshall 2015). These factors seem to set a sound starting point to inspire further research.

3 Research and Methodological Approach

Nowadays, school buses are seen as an innovative, sustainable alternative in urban Following the above considerations, conceptual framing and state of the art on the subject of territorial rankings a concern emerged: Are citizens represented and participating in city rankings? What is their role in the construction of city rankings narratives?

To find possible answers to this main research question, we set the following objectives: **to map scoring and categories** in city rankings; **to identify eventual citizen representational and/or participatory criteria**; **to critically reflect on** city rankings **purposes, meanings and objectives**.

The methodology used is qualitative and relies on hermeneutical interpretation and content analysis of three recognised city rankings: **City Brands Index 2017 (CBI), 2018 Global Cities Report (GCR)**, and **Global Liveability Index 2018 (GLI).**

The choice of these rankings was determined by notoriety on the Google research engine, but also by updating, access, availability and relevance. CBI – City Brands Index 2017 is an iconic city ranking, heir of the first incursions in territorial communication models, GCR – 2018 Global Cities Report, was selected for its broad scope and Global Liveability Index 2018 in order to include liveability criteria to the research, in an effort to eventually better representing citizens' insights on the cities.

The content analysis focused on the explicit categories published by each ranking to frame the data collected. Methodological explanation and transparency was also taken into account.

4   Results and Discussion

The three selected city rankings content was scrutinized in order to comparatively assess: the scope and methodology used; the cities that reached the top 10 scores, the categories proposed by these rankings to assemble data in a meaning making narrative, and the representation/participation of citizens.

4.1   The Rankings: Scope and Methodology

The CBI – City Brands Index 2017 is also identified as the "Anholt-GfK City Brands IndexSM (CBISM)". This index inherits the knowledge work ground of Simon Anholt's—both the Nation Brands Index and the City Brands Index were pioneering in 2005, "is run once every two years and measures the image of 50 cities" (Kirchener 2018, p. 3) and is published by GfK – Growth from Knowledge, a consultancy and market research company. Data was collected from 20 September to 3 October 2017 through "5,057 interviews across 10 countries (Australia, Brazil, China, France, Germany, India, Russia, South Korea, UK, and USA), with at least 500 interviews per country (…) The achieved sample in each country is weighted to reflect key demographic characteristics including age, gender, and education of the online population in that country" (Kirchener 2018, p. 3).

Six key dimensions are taken into account to rank cities in this index: "Presence (the city's international status and standing); Place (its physical outdoors aspect and transport); Prerequisites (basic requirements, such as affordable accommodations and the standard of public amenities); People (friendliness, cultural diversity, how safe one feels); Pulse (interesting things to do); Potential (the economic and educational opportunities available)" (Kirchener 2018, p. 2).

Results indicate the following ranking for the first 10 places: Paris, London, Sydney, New York, Los Angeles, Rome, Melbourne, Amsterdam, San Francisco and Berlin (Table 2). Nevertheless, links were provided to further contacts and information about the CBI – City Brands Index 2017, the index itself was not available, and the information provided was public only through press releases (Kirchener, 2018). People's insights (from 10 countries) are taken into account.

The GCR – 2018 Global Cities Report, published by A.T. Kearny, an American consultancy firm, "highlights regional trends revealed over the past decade and examines what defines the world's most influential cities" (Peterson et al. 2018, p. 1). It was established in 2008 assessing the performance of 60 cities. Based on the accumulated experience it position itself as going "beyond ranking cities (...) reveals which global cities are primed for urban transformation and growth" (Peterson et al. 2018, p. 1), exposing trends that indicate that nevertheless business activity, economics, information exchange and governance remain important criteria to score some points in the ranking, "human capital and cultural experience are also significant drivers of growth" (Peterson et al. 2018, p. 2).

The results are divided in two comparative strengths reports: the Global Cities Index, based on performance, and the Global Cities Outlook, based on potential. For this research we will take into account exclusively the Global Cities Index, centred in the current performance of cities.

This index analyses 135 cities worldwide, covering Africa (13), Asia Pacific (44), Europe (24), Latin America (15), North America (16) and Middle East (13). 27 variables grouped into a set of 5 weighted categories are considered to narrow the ranking score: "Business activity (capital flow market dynamics and major companies present); Human capital (education levels); Information exchange (access to information through internet and other media sources); Cultural experience (access to major sport events, museums, and other expos); Political engagement (political events, think tanks, and embassies)" (Peterson et al. 2018, p. 12). The first 10 cities (Table 2) in this ranking are: New York, London, Paris, Tokyo, Hong Kong, Los Angeles, Singapore, Chicago, Beijing, Brussels (Peterson et al. 2018, p. 13).

Methodology is explained in rather brief traits, more centred on how information is assembled and organised than on how it is actually collected, stating that "sources are derived from publicly available data" (Peterson et al. 2018, p. 12) and eventually substituted by country data if city data is not available.

The Global Liveability Index 2018, published by The Economist Intelligence Unit, states: "The concept of liveability is simple: it assesses which locations around the world provide the best or the worst living conditions. Assessing liveability has a broad range of uses, from benchmarking perceptions of development levels to assigning a hardship allowance as part of expatriate relocation packages" (EIU 2018, p. 7).

To find the most liveable cities of the world The Economist team analysed a number of complex variables (30) of 140 cities and aggregated them in 5 major metrics: stability, healthcare, culture and environment, education, and infrastructure (Abadi 2018).

Methodological processes are available and an explanatory chapter exposes how the rating is calculated, "For quantitative variables, a rating is calculated based on the relative performance of a location using external data sources" (EIU 2018, p. 8). However, the fact that "For qualitative variables, an "EIU rating" is awarded based on the judgment of in-house expert country analysts and a field correspondent based in each city" (EIU 2018, p. 8) reveals that the qualitative insight of this index is subject to personal sensibilities, however informed they may be, which may rise some concern over the credibility and impartiality of this kind of value judgment.

According to this index, the 10 most liveable cities in 2018 were Vienna, Melbourne, Osaka, Calgary, Sidney, Vancouver, Toronto, Tokyo, Copenhagen and Adelaide (Table 2).

Not surprisingly, some conclusions of the report account for "a correlation between the types of cities that sit at the very top of the ranking. Those that score best tend to be mid-sized cities in wealthier countries. Several cities in the top ten also have relatively low population density (…) foster a range of recreational activities without leading to high crime levels or overburdened infrastructure" (EIU 2018, p. 4). In parallel the index draws attention to the victims of success, "The "big city buzz" that they enjoy can overstretch infrastructure and cause higher crime rates. New York (57th), London (48th) and Paris (19th) are all prestigious hubs with a wealth of recreational activities, but all suffer from higher levels of crime, congestion and public transport problems than are deemed comfortable" (EIU 2018, p. 4). Nevertheless they seem to keep their attraction factor. "The question is how much wages, the cost of living and personal taste for a location can offset liveability factors" (EIU 2018, p. 4).

Comparatively, GCR – Global Cities Report 2018 (GCR) and The Global Liveability Index 2018 (GLI) cover an approximate number of cities worldwide, while CBI – City Brands Index 2017 (CBI) covers only 50 cities (Table 1). However, CBI is based on a qualitative assessment through online interviews intended to assess "the image" of the cities, whether GCR is grounded on publicly available data on 27 metrics and GLI uses a mixed methodology, based on quantitative data and qualitative assessment of experts (Table 1).

Table 1. City rankings' scope/method comparative analysis.

Publisher	Scope	Method
CBI – City Brands Index 2017		
Anholt-GfK	50 cities	5,057 online interviews in 10 countries; sample to reflect demographic reality
GCR – Global Cities Report 2018		
ATKearny	135 cities	Publicly available city data; 27 metrics;
The Global Liveability Index 2018		
The Economist Intelligent Unit	140 cities	30 qualitative and quantitative metrics

4.2 The Cities

An overview of the cities ranked in the top 10 positions in the different rankings analysed in this research reflect a diversity of assessments, exposing that the best city images, the best city performances and the most liveable cities in the world do not necessarily match (Table 2).

Actually, there is not any city that occupies one of the 10 top positions in all the three rankings. The best possible match appears to be London (number 2 in two rankings); present in two rankings but in diverse positions stand six cities (Los Angeles,

Melbourne, New York, Paris, Sydney, Tokyo); and present in only one ranking are 16 cities (Adelaide, Amsterdam, Beijing, Berlin, Brussels, Calgary, Chicago, Copenhagen, Hong Kong, Osaka, Rome, San Francisco, Singapore, Toronto, Vancouver, Vienna).

There seems to be a prevalence of big cities, geographically concentrated in Europe (8), North America (7) and Asia-Pacific (7).

Table 2. City rankings' top 10 cities comparative analysis

2017 CBI – City Brands Index	2018 GCR – Global Cities Report	The Global Liveability Index 2018
1. Paris	1. New York	1. Vienna
2. London	2. London	2. Melbourne
3. Sydney	3. Paris	3. Osaka
4. New York	4. Tokyo	4. Calgary
5. Los Angeles	5. Hong Kong	5. Sidney
6. Rome	6. Los Angeles	6. Vancouver
7. Melbourne	7. Singapore	7. Toronto
8. Amsterdam	8. Chicago	8. Tokyo
9. San Francisco	9. Beijing	9. Copenhagen
10. Berlin	10. Brussels	10. Adelaide

Concerning city image (CBI), Europe gets the most (5), followed by North America (3) and Asia-Pacific (2). The most performative cities (GCR) are mainly in Asia-Pacific (4), followed ex-aequo by Europe (3) and North America (3). Regarding the best cities to live (GLI), Asia-Pacific seems to offer more (5), followed by North America (3) and Europe (2).

4.3 The Categories

Analysing comparatively the city rankings' content is an attempt to find patterns that might produce further knowledge. Therefore, in this case, we will try to categorise the city rankings' categories.

A general overview portrays an apparent mismatch of the three city rankings at stake, pulverized in a diversified set of categories (Table 3).

CBI presents data on the image of 50 cities aggregated in presence, place, pre-requisites, people, pulse and potential. Since it is based on interviews it reflects the perceptions people have on cities, whether they live there, visited the cities or not. To a certain extend it is a ranking of impressions that were constructed by direct experience, by the media, by cultural acknowledgement of the place and forged through accumulated reputational capital.

GCR reflects a mix of business activity, human capital, information exchange, cultural experience, and political engagement, whereas GLI categorises cities over stability, healthcare, culture and environment, education, infrastructure.

The first layer of analysis exposes two sets of categories: tangible—such as "business activity", "infrastructure" or "information exchange"—and intangible—such as "pulse" or "potential".

In a second layer we can find some common concepts or words that might qualify as categories. For example: culture, education, economics, environment, health or politics. They all seem important factors to assess a city.

The analysis gets more complex when we try to understand more deeply the semantics of these categories. What does "stability", "pulse" or "political engagement" mean? A "steady job"? The "vibe of economic growth"? The "activist output" of a city? According to the research, stability is the "prevalence of petty and/or violent crime; threat of terror, military conflict and/or of civil unrest/conflict"(GLI); pulse are "interesting things to do" (CBI) on a city and political engagement includes "political events, think tanks, and embassies" (GCR).

The examples above, exploratory as they are, expose the assumption that although categories are useful to aggregate data and produce a narrative, they but might be misleading to understand data and, without proper contextualisation, may mean literally anything.

Categories are a form of simplification and aggregation of complex data in order to make it intelligible, usable and meaningful. From all parameters of the present research this is the one where the strategic intentions of the city ranking producers might be more evident, as setting categories is an immediate form of interpreting reality, and therefore an attempt to control it.

Categories and the selection of criteria they entail are also indexed to the agenda setting of city rankings production. The case of GLI is quite illustrative. Published by The Economist Intelligent Unit is promoted as a Global Liveability City Index but was designed as a tool "to help companies decide how much "hardship" allowance they would need to pay employees who relocate (…) focuses on things that matter to expats, not citizens" (Rozek et al. 2018). But this is not a hidden agenda. The GLI explains that "assessing liveability has a broad range of uses, from benchmarking perceptions of development levels to assigning a hardship allowance as part of expatriate relocation packages (…) quantifies the challenges that might be presented to an individual's lifestyle in any given location, and allows for direct comparison between locations" (EIU 2018, p. 7). Nevertheless, widespread short versions of the city index, namely as catchy titles in media about new winners and losers (McArthur and Robin 2019), frequently make this agenda invisible, adding up to other critics, on lack of transparency: "It appears that, beyond the well-designed league tables and flurry of media attention, The Economist's Global Liveability Index is a mostly subjective rating with opaque methods for comparing cities (Rozek et al. 2018).

Therefore, to fully understand what a city ranking stands for, it is crucial to go deep into strategic and methodological explanations (when and if they are available).

4.4   The Citizen Representation/Participation

The issue of citizen representation and participation poses methodological challenges. Not only due to the above-mentioned difficulties on tracking data collecting processes and sources, but also due to the fact that citizens are embedded in the notion of the city.

Table 3. City rankings' categories comparative analysis.

Categories

CBI – City Brands Index 2017

- Presence (the city's international status and standing);
- Place (its physical outdoors aspect and transport);
- Pre-requisites (basic requirements, such as affordable accommodations and the standard of public amenities);
- People (friendliness, cultural diversity, how safe one feels);
- Pulse (interesting things to do);
- Potential (the economic and educational opportunities available)

GCR – Global Cities Report 2018

- Business activity (capital flow market dynamics and major companies present)
- Human capital (education levels)
- Information exchange (access to information through internet and other media sources)
- Cultural experience (access to major sport events, museums, and other expos)
- Political engagement (political events, think tanks, and embassies)

The Global Liveability Index 2018

- Stability (prevalence of petty and/or violent crime; threat of terror, military conflict and/or of civil unrest/conflict)
- Healthcare (availability and quality of private/public healthcare; availability of over-the-counter drugs; general healthcare indicators)
- Culture and environment (humidity/temperature rating; discomfort of climate to travellers; level of corruption and censorship; social or religious restrictions; sporting and cultural availability; food & drink; consumer goods & services
- Education (availability and quality of private education; public education indicators)
- Infrastructure (quality of road network, public transport; international links; energy provision; energy and water provision; and telecommunications; availability of good quality housing)

In the comparative analysis of the three city indexes is possible to observe indirect and direct representation and participation of citizens.

We would consider indirect representation/participation are present in variables related with citizens' actions and behaviours that contribute to construct the state of the city, such as "prevalence of petty and/or violent crime" (GLI); "level of corruption and censorship" (GLI) or "friendliness" (CBI).

Direct citizen representation/participation can be observed in the instrumental contribution of interviewer's opinions in the construction of city rankings, as in the case of CBI – City Brands Index. However, this direct participation is an indirect representation, if representational at all, because in the case of CBI the interviews were applied in 10 countries to assess the image of 50 cities, that do not necessarily coincide with a direct, living experience of the cities included in the ranking.

5 Conclusion

Throughout this reflective exploratory research, we aimed at finding whether city rankings, as assessing and promoting territorial communication tools, take the citizens into account, embedded in the notion of a smart city made by and for the citizens, grounded on communal capital and collective knowledge.

Conclusions point out that, **as city representation, city rankings prove to be reductive, frequently misleading,** and—although based on citizen status statistics, often collected from publicly available data, **tend not to embed citizen participation and insights**, thus exposing a representational and participatory gap. We would argue that, in order to effectively represent cities, as they are lived, thought and built by their citizens in their everyday, city rankings must rely in more real-time, updated, people's perception centred data, and embed more citizen participation and insights.

Although generally accepted and widespread through the media as assessing reports of the state of cities, city ranking's value seems to be strategically communicational and enhanced by its news worthy value. City rankings are (and set) opportunity to communicate, therefore can be seen and used more because of their value as promotional tools rather than because of their scientific input relevance.

This does not necessarily question sound scientific procedures in data collection of reputed organisations that contribute to knowledge. Actually, methodological processes within this research were briefly explained but not fully disclosed. Nor was this the main focus of the investigation. Nevertheless, more transparency and accountability of city ranking promoters might add to trust and credibility and more trans-disciplinary teams might enrich the insight and scope of city rankings.

Rather paradoxically, it seems to be not mainly the methodological design, but the use that is made of city rankings that puts city rankings at stake.

It is commonly accepted that city rankings contribute to construct meaning and add value to the cities, enhancing performance and competition. Notwithstanding we found that incomparable, out-dated or poor data that is simplified for the sake of understanding, measuring and controlling territorial realities, serves frequently as ground for assessment, thus establishing a rhetorical construct that does not mirror reality but a propagated through the media and commonly accepted narrative. Categories are a form of simplification and aggregation of complex data in order to make it intelligible, usable and meaningful. From all parameters of the present research this is the one where the strategic intentions of the city ranking producers might be more evident, as setting categories is an immediate form of interpreting reality, and therefore an attempt to control it, not necessarily to assess it.

Furthermore, the political economy of city rankings must be considered when assessing city rankings. Studying data costs money and the three cases we studied are designed and promoted by private organizations, however reputed, with their own agenda. Therefore, as powerful medicines, city rankings should be used with care and critical insight.

A final note on the limitations of the present reflection, based on an exploratory research: we believe further and larger scope investigation might produce different results. Next steps in research would go through more specific rankings and criteria, eventually to direct citizen perceptions—such as liveability, sustainability or happiness—and participation—such as activism, political involvement, direct deliberation. Both directions might produce a city ranking that reflects a city as it is lived and idealised by the citizens.

References

Abadi, M.: The 10 most livable cities in the world in 2018, 25 August 2018. Business Insider. https://www.businessinsider.nl/most-livable-cities-in-the-world-2018-8/

Anholt, S.: Competitive Identity: The New Brand Management for Nations, Cities and Regions. Palgrave Macmillan, Basingstoke (2007)

Atlee, T.: Participatory Sustainability – Notes for an Emerging Field of Civilizational Engagement. CreateSpace, Co-Inteligence Institute, Eugene (2017)

Atlee, T.: The Tao of Democracy: Using Co-Intelligence to Create a World That Works for All. North Atlantic Books, Berkeley (2014)

Atlee, T.: Empowering Public Wisdom: A Practical Vision of Citizen-Led Politics, vol. 4. North Atlantic Books, Berkeley (2012)

Buhmann, A., Ingenhoff, D., White, C., Kiousis, S.: Charting the landscape in research on country image, reputation, brand, and identity. In: Ingenhoff, D., White, C., Buhmann, A., Kiousis, S. (eds.) Bridging Disciplinary Perspectives of Country Image, Reputation, Brand, and Identity, pp. 1–10. Routledge, New York (2019)

Cabannes, Y.: Another City is Possible with Participatory Budgeting. University of Chicago Press Economics Books, Chicago (2019)

https://placebrandobserver.com/smart-cities-how-ict-impacts-urban-development-representation/

Cardullo, P., Kitchin, R.: Being a 'citizen' in the smart city: up and down the scaffold of smart citizen participation in Dublin, Ireland. GeoJournal 84(1), 1–13 (2019). https://doi.org/10.1007/s10708-018-9845-8

Casanova, M., Renner, M., Bielenstein, T.: Ranking of the Rankings, Rating the Rankings, in Branding Institute CMR homepage (2018) https://www.branding-institute.com/rating-the-rankings/ranking-of-the-rankings. Accessed 30 July 2019

Chourabi, H., et al.: Understanding smart cities: an integrative framework. In: 45th Hawaii International Conference on System Sciences, 4–7 January 2012. IEEE (2012)

Dameri, R.P.: Searching for smart city definition: a comprehensive proposal. Int. J. Comput. Technol. 11(5), 2544–2551 (2013)

Drucker, S., Gumpert, G.: The communicative City Redux, international. J. Commun. 10(2016), 1366–1387 (2016)

EIU: The Global Liveability Index 2018 A free overview, Economist Intelligence Unit (2018). http://www.eiu.com/Handlers/WhitepaperHandler.ashx?fi=The_Global_Liveability_index_2018.pdf&mode=wp&campaignid=Liveability2018. Accessed 12 Aug 2019

Giffinger, R., Haindlmaier, G., Kramar, H.: The role of rankings in growing city competition. Urban Res. Pract. 3(3), 299–312 (2010)

Hollands, R.G.: Will the real smart city please stand up? Intelligent, progressive or entrepreneurial? City 12(3), 303–320 (2008). https://doi.org/10.1080/13604810802479126

Ingenhoff, D., White, C., Buhmann, A., Kiousis, S. (eds.): Bridging Disciplinary Perspectives of Country Image, Reputation, Brand, and Identity. Routledge, New York (2019)

Kavaratzis, M.: City Branding. In: The Wiley Blackwell Encyclopedia of Urban and Regional Studies, pp. 1–4 (2019)

Kavaratzis, M., Kalandides, A.: Rethinking the place brand: the interactive formation of place brands and the role of participatory place branding. Environ. Plan. A 47(6), 1368–1382 (2015)

Kirchener, C.: Paris continues its reign as highest-rated city, Press Release, 30 January 2018. Anholt-GfK (2018). https://www.gfk.com/fileadmin/user_upload/dyna_content/Global/documents/Press_Releases/2018/20180130_PM_CBI_2017_efin.pdf. Accessed 3 Aug 2019

Maddox, T.: Smart cities: a cheat sheet, TechRepublic, Innovation, 16 July 2018. https://www.techrepublic.com/article/smart-cities-the-smart-persons-guide/. Accessed 3 Aug 2019

Marsal-Llacuna, M.L., Colomer-Llinàs, J., Meléndez-Frigola J.: Lessons in urban monitoring taken from sustainable and livable cities to better address the Smart Cities initiative. Technol. Forecast. Soc. Chang. 90(Part B), 611–622 (2015). https://www.sciencedirect.com/sdfe/reader/pii/S0040162514000456/pdf. Accessed 16 Aug 2019 (2015)

Marshall, A.: The Top 6 Reasons to Be Wary of City Rankings, Ranked, CityLab 20 May 2015 (2015). https://www.citylab.com/life/2015/05/the-top-6-reasons-to-be-wary-of-city-rankings/393686/. Accessed 14 Aug 2019

Martin, C.J., Evans, J., Karvonen, A.: Smart and sustainable? Five tensions in the visions and practices of the smart-sustainable city in Europe and North America. Technol. Forecast. Soc. Chang. 133, 269–278 (2018)

McArthur, J., Robin, E.: Victims of their own (definition of) success: urban discourse and expert knowledge production in the Liveable City. Urban Stud. 56(9), 1711–1728 (2019)

Meijering, J.V., Kern, K., Tobi, H.: Identifying the methodological characteristics of European green city rankings. Ecol. Indic 43, 132–142 (2014)

Melo, A.D.: A Marca dos Cidadãos: Intervenção, Ocupação e Subversão na Paisagem Publicitária Urbana. In: Pires, H., Mesquita, F. (eds.) Publi-Cidade e Comunicação visual urbana, pp. 61–89. CECS, Braga (2018)

Melo, A.D.: "Somos todos smart?": co-inteligência, comunicação e sustentabilidade territorial. In: Abreu, J. (ed.) Inteligência Territorial - Governança, Sustentabilidade e Transparência, Idioteque (2019)

Merkelsen, H., Rasmussen, R.K.: Evaluation of nation brand indexes. In: Ingenhoff, D., White, C., Buhmann, A., Kiousis, S. (eds.) Bridging Disciplinary Perspectives of Country Image, Reputation, Brand, and Identity, pp. 69-84. Routledge, New York (2019)

Newburry, W., Song, M.: Nation branding, product-country images, and country rankings. In: Ingenhoff, D., White, C., Buhmann, A., Kiousis, S. (eds.) Bridging Disciplinary Perspectives of Country Image, Reputation, Brand, and Identity, pp. 49–68. Routledge, New York (2019)

O' Sullivan, F.: Death to Livability! CityLab, 26 June 2019 (2019). https://www.citylab.com/life/2019/06/best-cities-to-live-list-monocle-ranking-zurich-vienna/592492/. Accessed 14 Aug 2019

Okulicz-Kozaryn, A.: City life: rankings (livability) versus perceptions (satisfaction). Soc. Indic. Res. 110(2), 433–451 (2013)

Papp-Váry, Á.: The Anholt-GMI city brand hexagon and the saffron European city brand barometer: a comparative study. Reg. Bus. Stud. 3(Suppl 1), 555–562 (2011)

Peterson, E.R., Hales, M., Peña, A.M., Dessibourg-Freer, N.: 2018 Global Cities Report - Learning from the East: Insights from China's Urban Success. A. T. Kearney (2018)

Rizzon, F., Bertelli, J., Matte, J., Graebin, R.E., Macke, J.: Smart City: um conceito em construção. Revista Metropolitana de Sustentabilidade 7(3), 123–142 (2017). ISSN 2318-3233

Rozek, J., Giles-Corti, B., Gunn, L.: The world's 'most liveable city' title isn't a measure of the things most of us actually care about, The Conversation Trust (UK) Limited, August 15, 2018 7.18am BST (2018). http://theconversation.com/the-worlds-most-liveable-city-title-isnt-a-measure-of-the-things-most-of-us-actually-care-about-101525. Accded 20 Aug 2019

Stremtan, F.: Some consideration regarding collective intelligence. In: 6th Annual International Conference of Territorial Intelligence, "Tools and Methods of Territorial Intelligence", p. 10, October 2008

Vanolo, A.: Smartmentality: the smart city as disciplinary strategy. Urban Stud. 51(5), 883–898 (2014)

Yigitcanlar, Tan, Inkinen, Tommi: Benchmarking City Performance. Geographies of Disruption, pp. 159–197. Springer, Cham (2019). https://doi.org/10.1007/978-3-030-03207-4_12

Zhang, F., et al.: Privacy-aware smart city: a case study in collaborative filtering recommender systems. J. Parallel Distrib. Comput. **127**, 145–159 (2019)

Mobility Time Style: For an Integrated View of Time and Mobility in Societies with a Future

Catarina Sales Oliveira[1,2] (ID) and Emília Araújo[3(✉)] (ID)

[1] University of Beira Interior, Covilhã, Portugal
catarinasalesoliveira@gmail.com
[2] Centre for Research and Studies in Sociology, University of Lisbon, Lisbon, Portugal
[3] Communication and Society Research Centre, University of Minho, Braga, Portugal
emiliarrajo@gmail.com

Abstract. This paper aims (i) to demonstrate the relevance of deepening the study of time usage from the individual and family perspectives and (ii) to put in dialog perceptions and uses of time with daily mobility patterns. It is increasingly imperative to consider mobility and the uses of time as central axes of lifestyles, highlighting the weight of several variables in the definition of lifestyle choices, namely transportation options. This reflection is based on an empirical study carried out in Portugal through interviews in the metropolitan areas of Lisbon and Porto. The analysis leads to the conclusion that, in addition to physical distances people have to cover, the choice of specific means of transportation is strongly dependent on the perceptions and uses of time. It is also evident that time is simultaneously dependent on the way technologies are absorbed into daily life and that time remains a matter of constraint and social opportunity.

Keywords: Mobility · Time · Styles · Lifestyle

1 Introduction

An increasing number of studies have given attention to the notion of quality of life [1] – especially wellbeing – when discussing mobility and travel within the context of modern cities [2, 3]. Most of these state that mobility practices and expectations have a strong impact on individuals and families' decisions and choices throughout their lives [2] conditioning the ways they use time and thereby being a component of people's lifestyles [4]. Authors [4] have claimed that the location of work activities is central in mobility as this strongly influences time use possibilities both in individual and collective terms and determines transportation means (type and forms of use). In this sense, several approaches to the city' spaces and times have fundamentally showed the effects of rationalization of both these dimensions. Later, in the context of globalization, migration and digitization of work further transformations in the form of organization and the spatial functioning of work emerge (each day increasingly more mobile and flexible) creating rather unpredictable configurations of uses of time and transportation.

© ICST Institute for Computer Sciences, Social Informatics and Telecommunications Engineering 2020
Published by Springer Nature Switzerland AG 2020. All Rights Reserved
P. Pereira et al. (Eds.): SC4Life 2019, LNICST 318, pp. 170–183, 2020.
https://doi.org/10.1007/978-3-030-45293-3_13

In Portugal, 42% of the national population lives in Lisbon and Porto metropolitan areas [5]. The development of these areas is intrinsically related to the process of urbanization: fuelled by the historic exodus from the inner country in the sixties of the last century, these cities are currently also experiencing a great wave of touristic influx. Presently, Lisbon hosts two million and eighty thousand active employees - about 27% of the national population [5]. Demographic changes happened in various locations around Lisbon over recent years. These are the cases of cities like Mafra, and Sesimbra which have grown exponentially. The growth in these places is an effect, among others, of the increasing use of private transport and the implementation of public infrastructures such as the river Tejo bridges [6]. This massive growth has gradually triggered the development of an urban continuum between Lisbon, Cascais, Sintra, Loures and Vila Franca de Xira on the north bank, and Sesimbra and Setúbal, on the south. Porto also shows a growth towards more distant places such as Póvoa de Varzim (north) and São João da Madeira (south) [7, 8].

This population growth into further places away from the cities is leading to increasing needs for transport, as the population needs to cover longer distances to access work, services, and other goods. Moreover, the lack of or the weakness of collective and public means of transportation, some of them reaching a rupture point (as happens with the boasts crossing the River Tejo) is leading to increasing use of private cars, which causes problems of traffic and congestion which negatively impact quality of life.

The questions of mobility are therefore complex insofar as poor political intervention concerning transportation solutions and living costs, together with increasing capitalist pressures over certain cities have consequences over people's individual lifestyles, impacting on quality of life. An integrated politics of space and time in the cities, perhaps by improving the technological facilities, would be a way of making cities available to all. An increasing number of studies have been giving attention to quality of life [5, 6] - especially wellbeing – when discussing mobility and travel relationship within the context of modern cities. Most of them state that mobility practices and expectations have a strong impact on individuals and families' decisions and choices throughout their lives [5, 6], the conditioning of the ways they use time thereby being a component of people's lifestyles.

This text explores the concepts of mobility and time within lifestyles, proposing the notion that increasingly, the analysis and the interventions on space – be they physical or digital – must include citizens' perceptions, values and uses of time. In fact, societies are increasingly digitally oriented, as more and more daily activities are accomplished digitally and have become, overall, ever more accelerated. However, citizens must individually deal with a multitude of different time constraints that are sociologically relevant, as they reveal unequal patterns of mobilities, as well as dissimilar accesses to means of mobility and technological trends. This, in course, leads to different uses of kinds of time or time styles (according to gender, age, social classes, and type of job), as well as different lifestyles, and mobility profiles.

2   The Importance of Mobility

There are manifold issues related to mobility and sustainability, particularly in the face of the problem of climate change and the development of adaptation and mitigation strategies that imply deepening the social processes of enormous relevance.

Mobility is a core issue for modern societies. Life choices are related to mobility behaviours and decisions, as each life choice seems to be linked to a specific set of mobility demands. As an example, decisions about work and home taken in a person's early twenties might not be desirable or feasible ten or twenty years later, when responding to family growth [9] or to the process of ageing. In a way, these decisions should all converge into a good life choice. In other words, if mobility has long been known as a feature of human groups, its expression in contemporary society takes on unexpected features that strongly impact society in different realms [10, 11]. International research on mobility has greatly increased in recent years.

Sheller and Urry came to the conclusion that mobility involves such different aspects of the individual and the social world that it actually can be understood as an analytical paradigm; the authors have clearly demonstrated its multidimensionality, thus taking into account the role of technology, cultural expressions and social expectations among other dimensions [12]. Eurostat reports indicate that in Portugal (alongside Norway) car usage is higher than that of the European average, representing 89.8% of the passenger transport [13–15]. The data and Eurostat [15] indicate that transport is one of the main sources of energy consumption in Europe and, as a result, private consumption reaches values close to consumption in industry. The transition roadmap for a competitive low-carbon economy in 2050 [15] stipulates a set of objectives to be achieved over the next 30 years regarding decarbonization, innovation and the adoption of green energy. But the data [16] also indicates that transport costs continue to increase, while the consumption of renewable energies in transport remains low [9]. These results match with the demonstrations of the IEA-International Energy Agency – according to which the consumption of biofuel in road transport will only increase by 1% in these 7 years (2015–2020)

Briefly, all this information helps to support the idea that, alongside any technical and/or scientific changes that involve mobility, displacement and the use of time, are variables and processes of cultural and social character that should be considered, either from the point of view of the positive effects they may cause, or looking to the consequences societies will have to deal with, in the near future, in order to build democratic access to mobility and transportation.

3   Mobilities, Lifestyles and Time Uses

The relation between lifestyles and mobility has been recently explored by Cohen et al. who propose that "lifestyles and forms of mobility increasingly co-mingle in ways that can be crucial to the lives of those who are privileged enough to access them" [4]. At its core, lifestyle is an intricate result of choices that encompasses space and time uses [17]. The connection between lifestyles and mobility supposes free choice and in that sense the concepts of mobilities, lifestyles, and hypermobility [4] are close to the concept of migration lifestyle. Lifestyle conceptualization implies considering the following elements/dimensions: job, housing, and family. It is true that jobs are no longer lifelong and moving (whether actually or virtually) is a main driving force in people's lives. However, the weight of these transformations introduces ever more variations on the patterns of the course of life [18] Therefore, these three usual milestones - work, housing and family - remain as structuring variables, even if only in a mid-term perspective.

Flamm and Kaufmann [19] have proposed that within a world which is becoming increasingly mobile, whilst some people have largely increased their potential for mobility or motility, other are strongly deprived of this ability that negatively affects their quality of life [20] with respect to this, the recent contribution of the concept of mobility justice is of interest. In sum, mobility and travel options cannot be disconnected from the complex web of choices and constraints that a life path is made of [18], being an integrant component of lifestyles. Additionally, behaviours concerning mobilities are initially linked to choices and constraints concerning patterns of using and experiencing time. Mobility requires discussions on time as it is a core element of a lifestyle [17, 21]. Time has been analysed in different scientific areas both as social object, as a resource and a source of meaning. Specifically, in sociology, besides the studies that addressed the connections between time and space [22, 23] researchers from different historical and sociological contexts state the relevance of time usage on people's lives. The assumption that working time is the dominant time that determines the feeling of fulfilment in other times of life, is a core topic of research, alongside increasing social acceleration and the ways it affects people's options and chances, impacting on all life spheres, including on mental and subjective wellbeing. More recently the sphere of work trends - increasing digitisation and automation, flexible working times, globalization and dislocation - have been responsible for the emergence of several new time orders and uses which affect, and in some cases collide with, family time, personal time, or leisure time. These changes have led to alterations in mobility patterns, including the means of transportations used. In fact, one of the reasons families often use the car and other private means of transportation is due to the accumulation of activities they must accomplish during the day, part of them with children and the need to accomplish these as fast as possible [24]. These changes have led to alterations in mobility patterns, including the means of transportations which are being used, and which are increasingly required to be more personalized and flexible. Extensive literature has analysed the expansion of the mobile society. Therefore, one of the pivotal reasons why families keep using the car is the accumulation of activities they must accomplish almost simultaneously along the day. This come to be especially acute in the cases when families must live far from workplaces, schools and health services, in response to the increase in housing prices.

For a general comprehension of the existing studies on mobility, we present the following Table 1:

Table 1. Some themes in the mobility studies and time uses [25]

Traditional themes	Emerging themes	Associated dilemmas and emerging issues
Transport network, timetables, routes and prices	Flexible collective transport Technological adaptation and transport use	Safety of children in parks, public spaces and dwellings
Network accessibility, timetables and circuits, population covered and excluded	Access and use of transport, and new social discriminations and exclusions	Noise pollution in housing areas, mental illnesses and school performance
Quality of services, stops and parking spaces	Speed, increased consumption and reduced rest time	Deadly and incapacitating road accidents
Security and accountability	Forced mobility, speed and fragmentation of family and personal times	Transportation of children, traffic intensity and security
Adequacy of infrastructures and signage adjusted to Mobilities, Lifestyles and choice of transports	Tourism, transport and circulation Energy transition, transport and social values	Speed, transport and new social, gender and age exclusions Transportation, lifestyles and aspirations (school choices and children's extracurricular activities)
The value and consumption of cars	Safety and road coexistence Space-time representations and cognitions	Mental illness in children and young people, panic and time pressure
Circulation, infrastructure and convenience	Virtual time space and driving styles	Accessibility to transport and social exclusion
	Security and social inequality	"Green" energy and social inequality in Access
	Sustainability, gender, climate change and energy	Energy transition, low carbon economy and new inequalities
		Flexibility in the transport network, notably as regards stops and socio-demographic profile of the public.

3.1 Bringing Together Mobility and Time to Understand Mobility Time Styles

In accordance with the above explanation, we can say that the concept of mobility time lifestyles involves three types of elements: present settlements and places of activity (house, work, leisure, family activities); time and duration of (in)activity (day/week/years; morning/afternoon/evening; working day/weekend/vacations, etc.);

and the motivations underlying these activities (inscribed in social and economic constraints and/or emerging from an individual decision). The concept allows overcoming the usual binary readings dealing with how these times and spaces interweave with each other in a certain configuration, while keeping flexibility and diversity in mind. To operationalize the concept, it is necessary to study two main dimensions: (i) the elements concerning what people do and how they do it with regards to time, space and mobility as well as what value - material and symbolic - they attribute to it (linked to subjective and objective class indicators); and (ii) the elements related to people's disposition to face decisions concerning time, space and mobility and its management (mostly linked to subjective class indicators, also inscribed in professional cultures) incorporating the dynamism or mobility of these decisions. Therefore, it is possible to consider that mobility and time lifestyles are result of the intercrossing between three main dimensions: (i) the degree of time-space standardization, (ii) the degree of time-space stability, and (iii) the degree of time-space (self-) coordination.

Each of those dimensions reflect a joint imbricated result between values and practices which imply time, space, and mobility issues.

This paper provides an analysis of two main kinds of mobility time, or "time styles" according to data collected from the study done in the cities of Lisbon and Porto, in Portugal.

4 Method

This research involved 31 in-depth interviews with people who live and work inside of the two Portuguese metropolitan areas of Lisbon and Porto. A purposive sample was employed, choosing people with different family profiles, and distinct working time regimes. The interviews were made throughout one year using an interview grid, which integrated questions intersecting time, space, mobility and lifestyles topics. The interviews were used to collect information about many dimensions concerning mobility and time uses. In this paper the objective is to use the information to stress the strength of argument calling for better understanding of the sociological and comprehensive aspects of time and space appropriations.

Therefore, information was analysed according to the aforesaid themes currently studied: (i) the degree of time-space standardization, (ii) the degree of time-space stability, and (iii) the degree of time-space (self-) coordination. Data about the interviewees socioeconomic profile, gender, age and other information is displayed in the annex.

5 Findings and Reflections

5.1 Profile 1: Liminal and Adjusted Mobility Time Lifestyle

The most relevant and shared feature of this profile emerging across interviews, is the way people mobilize strategies to cope with demands without having to break social rules or give up specific roles/desires. This profile therefore combines high levels of standardization with high levels of desire for stability, and in turn, lower levels of self-coordination of time and space. This style prevails among those interviewees with small children and other dependents, those who undertake more rigid routines. Here, routines play a fundamental role, as it becomes mandatory to have rigidly observed daily plans, linked to the schedules and mobility needs of other family members (children, spouse, etc.). For this section of the population, diaries and calendars are central, so that daily routines can be defined in advance. Therefore, the people exhibiting this profile type seek to control time and travel, and to manage these in a stable way. Whether using public or private transportation, this section of the population needs to have control over space and time (and distance) to ascertain their lives. The following quote is an account by an interviewee relating how predictability allows her to conciliate work and family spheres:

"Compared with the past, when I was always arriving home after 10pm, my current life is better, less stressed. I was [working] in the commercial business back then. [Now] I feel [I have] a shortage of time but that's no big deal, it's normal, everyone would like to have more things done. Now, with this [current] job, I have a stable routine, and I find this is very good and necessary when one has children. I was lucky to have always had a good support network." (Dianna, 38, mother of one, economist)

The following excerpt highlights how daily time is deeply dependent on children's mobility and time needs, in a rigid manner:

"I synchronize my time with that of my two children, 18 and 21 years old [boys], who study Law and History in Lisbon. We always come to and from the station together; sometimes we need to pick the eldest one up at 9pm from the station since there is no decent public transportation to take him home. Our household spends three [daily] hours in transit. I sleep very little." (Kate, 42, mother of two, lawyer)

This prevalent idea about the need to interconnect different mobility needs in a stable way within the same negotiations with time inside the family sphere appears to have a correspondence with individual and family needs at certain moments of their lives as well as with the way people look to their future needs.

For this group, having a car is central as it ensures the possibility of being mobile and arriving at required destinations rapidly. In almost all cases, anxiety and feelings of time pressure come associated with mobility needs. At first sight, this mobility time style can be associated with a low level of individual time-space coordination and a high level of stability, in the sense that people tend to search for steadiness, trying hard to have and maintain a life with a certain degree of stability. They tend to apply space-time discipline also to biographical paths, as they decide whether to undertake, anticipate, or postpone events, considering, in their decisions, the very nature of family members' mobility needs and prospects. In so doing, they also prioritize needs and desires on the scale of their lifetime. The next excerpt provides evidence of the high degree of importance that a time-space-mobility routine may have, thereby reinforce how crucial it can be to structure mobility time styles:

"For sixteen years now that we spend vacations in the same place, at the same time, in August. This year will be the same. I like it very much." (Kevin, 58, father of one, rail infrastructure manager).

5.2   Profile 2: Liminal and Extempore Mobility Time Lifestyle

This profile differs from the previous one as it combines high levels of self-coordination of time and space with elevated levels of stability, even when the individual is faced with high levels of unpredictability. This profile refers to individuals and families that reveal a great need to introduce levels of unexpected activities during the day, without being strict about the need to respect a certain time guideline. There are moments of time/space that can be interchanged in all spheres of life including domestic chores. What distinguishes those who match this style is the wiliness to develop actions without a focus in their planning. In some cases, there is some level of disorganization, while in others it is mainly a matter of flexibility and adaptability. This profile reflects the predominance of flexible working hours with a need to be mobile and often to adapt time demands. Nevertheless, there were also people with flexible working hours within the previous profile, however what differs is their attitude towards this flexibility. Below is a quote of one of the interviewees who shares this mobility time style:

"There is no consistency in my time, although there is a seasonal pattern at work, but there is [consistency] in [some of] my daily activities. I have breakfast relatively early at home, I make calls while I'm in route, I have lunch at the same place [every day], etc. I do not have a regular weekly schedule; I work more frequently during weekends. The concept of free time is something that I do not use because my professional and private lives are very mixed; I am always trying to enjoy holidays [while] combining business and social relationships [which] are in the same group." (David, 58, businessman)

Interestingly, people within this profile share one same distinctive feature: they work in arears that demand flexible time-space arrangements. Much more than those who belong to the previous group, these people also relate difficulties in planning their lives, and their work is consequently thought to be potentially less stable. Thus, they also show less availability to plan or anticipate future choices concerning their mobility prospect. One should note, however, that although some of these interviewees advocate time-space mobility, which is secured by the intense use of cars, they express the willingness to have the possibility to better anticipate their life choices and to be able to plan such choices in a medium time range. In other words, much of the interviewees in this profile would strongly appreciate having more ability to think and plan the future, even though they have different degrees of autonomy to coordinate and master time and space movements. Here, it is relevant to point out that most interviewees are conscientious of the degree of precariousness as well as uncertainty that permeate their lives' choices (mostly those working in the private sector, without working contracts, divorced, and/or with dependent small children).

6   Discussion

The two profiles differ with regards to sociodemographic characteristics.

The first group is composed mainly of married people with small children, employed in dependent work and using both public transportation and private cars.

The second is mainly composed of married people without dependents, working in areas that demand greater flexibility of time. Compared to the former group, these individuals use private cars more intensively.

There are two important ideas to retain: First, mobility as expectancy is mainly ascribed to the second group, as these persons combine mobility needs with professional demands and leisure activities, perceiving cars as the taken for granted means of private transportation. In the case of the first profile, this includes individuals who understand mobility mostly as a determined and defined time-space. For them, most of the actions involving mobility must be prepared according to other schedules, especially if using public transportation.

Additionally, this paper can conclude that the greater distinction between the two groups is not connected to professional activity and its degree of flexibility, but rather with the stage in which individuals are in terms of the position within their life cycle as well as with the existence of small children in the family (who need to be cared for).

We can therefore state that a mobility time style is a complex web of time-space arrangements, which are sculpted according to a set of subjective and objective variables. Such variables also encompass a complex combination of objectivized behaviours (such as car usage, the amount of time allocated to each activity or the distance travelled) and implicit and subjective behaviours that may vary along a life trajectory (plans, projects, dreams, among others). The way people think time and relate to it – in a more or less flexible manner – is crucial to understand to what extent they are also able to alter their mobility habits.

7 Concluding Remarks

This paper has presented some of the motives for which mobility diagnosis is increasingly important to understand the inner difficulties that characterize people's lives concerning decisions of where to live, and how to move to have access to all desired/required the services and goods. In societies such as the Portuguese one, where the consumption of cars is increasingly higher, studies about mobility and time use patterns are ever more needed, to rethink transportation, as well as several other spheres of life, such as work, schools, hospital and health services, culture and consumption in shops [10].

In order to conclude this brief contribution, it is important to remind ourselves of the need for local authorities to involve themselves systematically and directly in the collection and processing of information about people's mobilities, associated with time uses and constraints. The construction of contemporary information databases on mobility flows, motivations and difficulties of the population, implying contact with diverse sources, in addition to individuals or families, thus constitutes a point of great importance, serving as an instrument to support future decision-making.

Acknowledgment. This work was supported by FCT, through the Strategic Financing of the R&D Unit UID/SOC/03126/2019.

Annex

See Tables 2 and 3

Table 2. Socio-demographic profile of people inserted in a liminal adjusted mobility time style

Age	Place of residence	Gender	Means of transportation	Occupation	Family	Work place
32	Montijo	W	Bus	Telecommunications technician	Couple with a baby	Lisbon
34	Setúbal	W	Train	Human Resources assistant	Isolated	Lisbon
29	Setúbal	W	Train	Roaming technician	Couple with a young child	Lisbon
42	Azeitão	W	Car and train	Solicitor (private company)	Divorced couple with two grown-up children	Lisbon
38	Palmela	W	Car and train	Economist	Couple with a teenage daughter	Lisbon
33	Sesimbra	W	Car	Administrative assistant and PhD student	Couple with a daughter of school age	Lisbon
36	Moita	M	underground, ferry and train	Secretary	Couple with a son of school age	Lisbon
56	Oeiras	M	Underground and train	Rail infrastructure manager	Couple with an adult son	Lisbon
30	Cascais	W	Car	Product Manager	Couple, expecting a child (six months of pregnancy)	Lisbon
32	Alverca	W	Train	Secretary	Couple with a baby and a young child	Lisbon

(*continued*)

Table 2. (*continued*)

Age	Place of residence	Gender	Means of transportation	Occupation	Family	Work place
24	Almada	W	Train	Call Center Operator	Isolated and pregnant (2 months of pregnancy)	Lisbon
34	Amadora	M	Car	Human Resources Assistant	Couple with two sons of school age	Lisbon
31	Sintra	W	Train	Commercial Assistant	Couple with no kids	Lisbon
48	Sintra	W	Car	Teacher	Couple with two adult sons	Loures
40	Loures	W	Car and train	Health and Safety technician	Couple	Lisbon
31	Mafra	M	Car	Computer technician	Couple with a small baby and a daughter of school age	Lisbon
36	Barreiro	W	Train or ferry	University lecturer	Couple with a son of school age	Lisbon
29	Alcochete	W	Car	Product Manager	Couple with two young children	Lisbon
35	Lisbon	F	Car	Product Manager	Couple with two children of school age	Lisbon
36	VF Xira	F	Train	Secretary	Couple with a son of school age and a baby daughter	Lisbon

(*continued*)

Table 2. (*continued*)

Age	Place of residence	Gender	Means of transportation	Occupation	Family	Work place
36	Entroncamento	M	Train	Portuguese Railway Company employee	Couple with a daughter of school age	Lisbon
38	Póvoa Varzim	W	Car	Accountant	Couple with a daughter of school age	Porto

Table 3. Socio-demographic profile of people inserted in a liminal extempore oriented mobility time style

Age	Place of residence	Gender	Means of transportation	Occupation	Family	Work place
31	Odivelas	M	Car	Salesperson	Divorced	AML
32	Póvoa Varzim	M	Car	Medical doctor	Isolated (his girlfriend lives in Lisbon)	Porto
32	Lisbon	W	Car	Human resources director	Isolated	Loures
42	VN Gaia	W	Car	Teacher	Couple	VN Gaia
26	Póvoa Varzim	W	Car	Call center operator	Unmarried couple	Porto
58	Azeitão	M	Car	Businessman	Couple	Setúbal
43	Maia	M	Car	Salesperson	Couple	Porto
39	Seixal	W	Car	Driver and Managing Partner of a transportation company	Divorced with a daughter of school age	Seixal
38	Lisbon	M	Car	Marketing Director	Couple with son of school age and a baby	Lisbon

References

1. Schwanen, T., Wang, D.: Well-being, context and everyday activities in space and time. Ann. Assoc. Am. Geogr. **104**(4), 833–851 (2014)
2. De Vos, J., Schwanen, T., Van Acker, V., Witlox, F.: Travel and subjective well-being: a focus on findings, methods and future research needs. Transp. Rev. **33**(4), 421–442 (2013)
3. Wang, D., He, S. (eds.): Mobility, Sociability and Well-being of Urban Living. Springer, Heidelberg (2016). https://doi.org/10.1007/978-3-662-48184-4
4. Cohen, S.A., Duncan, T., Thulemark, M.: Lifestyle mobilities: the crossroads of travel, leisure and migration. Mobilities **10**(1), 155–172 (2015)
5. Pordata: Base de Dados Portugal Contemporâneo. https://www.pordata.pt/. Accessed 01 July 2019
6. Melo, C., Vala, F.: Movimentos pendulares e organização do território metropolitano: distâncias e proximidades nos sistemas metropolitanos de Lisboa e Porto. Revista Portuguesa de Estudos Regionais **5**, 5–34 (2004)
7. Portas, N., et al.: Políticas Urbanas II: transformações, regulação e projectos. FCG, Lisboa (2013)
8. Instituto Nacional de Estatística Proporção de utilização do automóvel nas deslocações (%) por local de residência (2011). https://www.ine.pt/xportal/xmain?xpid=INE&xpgid=ine_indicadores&indOcorrCod=0007135&contexto=bd&selTab=tab2. Access 01 July 2019
9. Cresswell, T.: Towards a politics of mobility. Environ. Plann. Soc. Space **28**(1), 17–31 (2010)
10. Zhang, J.: Revisiting residential self-selection issues: a life-oriented approach. J. Transp. Land Use **7**(3), 29–45 (2014)
11. Montulet, B., Hubert, M., Huynen, P.: Etre Mobile. Vécus du temps et usages des modes de transport à Bruxelles. FUSL, Bruxelles (2007)
12. Sheller, M., Urry, J.: The new mobilities paradigm. Environ. Plann. A **38**(2), 207–226 (2006)
13. Eurostat: Estatísticas sobre o Transporte de Passageiros. http://ec.europa.eu/eurostat/statistics-explained/index.php/Passenger_transport_statistics/pt. Accessed 01 July 2019
14. Comissão Europeia: Roteiro de Transição para uma Economia Hipocarbónica Competitiva em 2050.:http://eur-lex.europa.eu/legal-content/PT/TXT/HTML/?uri=CELEX:52011DC0112&from=EN. Accessed 01 July 2019
15. Eurostat: Consumption of Energy. http://eur-lex.europa.eu/legal-content/PT/TXT/HTML/?uri=CELEX:52011DC0112&from=EN. Accessed July 2019
16. Eurostat: Renewable Energy Statistics. http://ec.europa.eu/eurostat/statistics-explained/index.php/Renewable_energy_statistics. Accessed 01 July 2019
17. Jensen, M.: Defining Lifestyle. Environ. Sci. **4**(2), 63–73 (2007)
18. Zhang, J.: The life-oriented approach and travel behavior research. In: 14th International Conference on Travel Behaviour Research, Windsor, UK, 19–23 July (2015)
19. Flamm, M., Kaufmann, V.: Operationalising the concept of motility: a qualitative study. Mobilities **1**(2), 167–189 (2006)
20. Lucas, K., Jones, P.: Social impacts and equity issues in transport: an introduction. J. Transp. Geogr. **21**, 1–3 (2012)
21. Cockerham, W., Rutten, A., Abel, T.: Conceptualizing contemporary health lifestyles: moving Beyond Weber. Sociol. Q. **38**(2), 321–342 (1997)
22. Hagerstrand, T.: Dynamic Allocation of Urban Space. Saxon House, Farnborough (1975)
23. Harvey, D.: The Condition of Postmodernity: An Enquiry into the Origins of Cultural Change. Blackwell, Oxford (1989)
24. Oliveira, C.S.: Mobilidades em português: paradigma, cultura e potencialidades. In: Araújo, E., Ribeiro, R., Andrade, P., Costa, R. (eds.) Viver emla mobilidade: rumo a novas culturas de tempo, espaço e distância, pp. 10–21. CECS, Braga (2018)

25. Araújo, E.: Questões de mobilidade, tempo e sustentabilidade. In: Araújo, E., Ribeiro, R., Andrade, R. Costa, R. (eds.) Viver em la mobilidade: rumo a novas culturas de tempo, espaço e distância, pp. 146–160. Centro de Estudos de Comunicação e Sociedade, Braga (2018)

Author Index

Printed in the United States
By Bookmasters

SpringerBriefs in Complexity

SpringerBriefs in Complexity are a series of slim high-quality publications encompassing the entire spectrum of complex systems science and technology. Featuring compact volumes of 50 to 125 pages (approximately 20,000–45,000), Briefs are shorter than a conventional book but longer than a journal article. Thus Briefs serve as timely, concise tools for students, researchers, and professionals.

Typical texts for publication might include:

- A snapshot review of the current state of a hot or emerging field
- A concise introduction to core concepts that students must understand in order to make independent contributions
- An extended research report giving more details and discussion than is possible in a conventional journal article,
- A manual describing underlying principles and best practices for an experimental or computational technique
- An essay exploring new ideas broader topics such as science and society

Briefs allow authors to present their ideas and readers to absorb them with minimal time investment. Briefs are published as part of Springer's eBook collection, with millions of users worldwide. In addition, Briefs are available, just like books, for individual print and electronic purchase. Briefs are characterized by fast, global electronic dissemination, straightforward publishing agreements, easy-to-use manuscript preparation and formatting guidelines, and expedited production schedules. We aim for publication 8–12 weeks after acceptance.

SpringerBriefs in Complexity are an integral part of the Springer Complexity publishing program. Proposals should be sent to the responsible Springer editors or to a member of the Springer Complexity editorial and program advisory board (springer.com/complexity).

More information about this series at http://www.springer.com/series/8907

Ana Teixeira de Melo

Performing Complexity: Building Foundations for the Practice of Complex Thinking

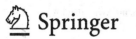 Springer

Ana Teixeira de Melo
Centre for Social Studies
University of Coimbra
Coimbra, Portugal

Ana Teixeira de Melo is funded by the portuguese Fundação para a Ciência e Tecnologia (FCT), under the transition norm of DL 57/2016 altered by Law 57/2017 (REF DL 57/2016/CP1341/CT0011).

ISSN 2191-5326   ISSN 2191-5334   (electronic)
SpringerBriefs in Complexity
ISBN 978-3-030-46244-4   ISBN 978-3-030-46245-1   (eBook)
https://doi.org/10.1007/978-3-030-46245-1

This Springer imprint is published by the registered company Springer Nature Switzerland AG
The registered company address is: Gewerbestrasse 11, 6330 Cham, Switzerland

To Leo and Susan

This work is dedicated to Leo Caves (my lifetime mindwalking partner) and Susan Stepney for their generosity in sharing their time and the complexity of their thinking with me in multiple enriching and delightful conversations which nurtured and scaffolded the development of my own thinking. The limitations and flaws of the work here presented are, however, to be attributed to me only.

Preface

This book is presented in a linear form as a succession of chapters and sub-sections. Nevertheless, the thinking that supported its development was not linear, nor was the process of writing. Each part was developed in relation to the others and as one progressed the others were subject to changes. It is likely that the understanding of its core ideas will emerge only from the interaction of its parts. Hence, it is possible that some parts will seem insufficiently clear until later ones bring further claritication. At some points, ideas that were previously presented are summarized as, on the one hand, they deserve to be revisited in light of more recent statements and, on the other, they help to strengthen, to clarify or to contextualize recent points. The reader may, nevertheless, have a feeling of "dejá read". We invite the reader to adopt an attitude of openness and to exercise some form of recursive thinking (a corollary of complex thinking as presented), asking for trust that—in its recursiveness—is small, but nevertheless important, nuances of the ideas presented will become more evident, as well as their implications. This work is not closed nor complete. It is intended to contribute to building foundations of what we hope will become a prolific domain for the interdisciplinary exploration of the idea of complex thinking across theoretical, empirical, and practice levels.

This brief is not a literature review. Nevertheless, it attempts to introduce the reader to some key ideas brought to the fore by Complexity Sciences and the challenges of thinking complexity. It explores a concept of complex thinking that, coined by Edgar Morin, has long been circulating in the literature. Though used (and perhaps abused) many times, the concept has not, to our knowledge, been significantly developed since Morin's early proposals, particularly in pragmatic terms.

The concept of complex thinking requires further development into a theoretically-informed but pragmatic framework where it illuminates the work of (i) researchers, exploring the mysteries and surprises of complexity, and trying to understand its underlying processes, (ii) common practitioners, dealing daily with the challenges of managing change in the real world, (iii) managers and policy-makers, facing difficult decisions in contexts of uncertainty and incomplete knowledge and (iv) people of all kinds, creating complexity by virtue of their existence (in themselves, in their relationships, and in the environment around them) but not always knowing how to change their trajectories or manage their own creations. We attempt to build the foundations for the further development of a pragmatic approach aimed at understanding and performing complexity. Most of the parts of the arguments presented here are not new; they have been explored before, with much more eloquence, elegance, and rigor than we do justice to here. Nevertheless, we hope that the way we bring these ideas together generates some meaningful and relevant novelty, opening a new field of interdisciplinary exploration for theory, research, and practice development, that brings together scientists, philosophers, artists, practitioners, and policy-makers, as well as different communities of artists and the rest of us in building a more complex but also better world.

Coimbra, Portugal   Ana Teixeira de Melo

Acknowledgements

Ana Teixeira de Melo is a researcher at the Centre for Social Studies, University of Coimbra, Portugal.

This work was supported by the Fundação para a Ciência e Tecnologia (FCT) [Foundation for Science and Technology, Portugal] through the Transition Norm of DL 57/2016 altered by Law 57/2017 (DL57/2016/CP1341/CT001).

Preliminary studies leading to this work were supported by a postdoctoral fellowship [SFRH/BPD/77781/2011].

I thank the York Cross-Disciplinary Centre for Systems Analysis (YCCSA-formerly York Centre for Complex Systems Analysis) and the Department of Computer Science of the University of York for hosting me as a visiting academic, on several occasions, between 2016 and 2018. The time spent at YCCSA has offered me the most extraordinary experiences of deep and rich interdisciplinarity which had an immense impact on the development of my thinking and of the work presented here. I am indebted for the warm welcome, openness and intellectual generosity of its resident staff, and the nurturing environment afforded by YCCSA for the incubation of ideas.

I am also thankful to the Centre for Social Studies for supporting this exchange.

Many of the ideas explored in this book were *seeded* during the time Leo Caves and I worked on "*(Gardening) gardening*" (Caves & Melo, 2018). The conversations we had around that manuscript have fed my interest and curiosity and led me to a deeper engagement with the notion of complex thinking, exploring other properties beyond the elements of relationality that we explored in that book chapter. The work presented here has also benefited from our exploratory adventures into practice with the first experiences of the "Complex Thinking Academy". In fact, theory and practice have been evolving together in the context of a mutually influencing, recursive, relationship. My research and practice as a family scientist and a family psychologist have also contributed to deepen and sharpen my theoretical reflections about complex thinking.

I am thankful to all the teams of practitioners who have accepted the challenge of exploring the practice of complex thinking with me, in the context of the development of a "Guide for Complex Case Conceptualization" applied to support the assessment and promotion of the potential for family change (flourishing, well-being, positive change, resilience). A special thanks to Adriana Dias, Andreia Neves, Dina Reis of the CAFAP [Centre for Parental Counselling and Family Support] of St. Casa da Misericórdia da Murtosa, Patrícia Calado and Ema Lopes, of the CAFAP of Associação Crescer Ser, Marinha Grande; Claúdia Fernandes, Vânia Lemos and Carina Terra of CAFAP Raio de Sol; and to Ana Vita, Catarina Gomes, Patrícia Almeida e Ana Luísa Almeida of CAFAP Quinta do Ribeiro. This amazing group of practitioners have long been a part of my research program and of my learning.

About This Book

Complexity Sciences have introduced us to a multidimensional world of extraordinary capacities that challenges our habitual logics and our capacity to understand and manage change. In the face of increasingly more complex challenges, humanity is, more than ever, pressed to embrace complexity, to understand its underlying processes and to build effective ways of managing them. But our understanding of the complexity of the world is related to the understanding of our own complexity and our role both as its products and producers. It is necessary to develop modes of thinking more congruent with complexity: capable not just of attending to but of *performing complexity.* Inspired by the properties of the complex world, we explore notions of complex thinking (first and second order or emergent) as practices that create conditions for the enactment of complexity at the level of our thinking, that expand our possibilities for action; we relate complex thinking to forms of abductive thinking. We ground the concept of complex thinking in a relational pragmatic worldview, where complexity is constructed as a relative and evolving concept and a relational process: both as a process and an outcome of a coupling relationship between an observer and the world. We present an operational proposal for the concept of complex thinking through the definition of a set of dimensions and properties that together build the foundations for the further development of a pragmatic framework and meta-methodological approach to performing complexity in attempting to understand and manage change in complex systems.

Contents

About the Author

Ana Teixeira de Melo is a researcher at the Centre for Social Studies of the University of Coimbra. She holds a Ph.D. in Psychology, speciality of Clinical Psychology by the University of Coimbra. She obtained her undergraduate degree in Psychology (Licenciatura–5 years) by the University of Porto with the pre-specialization in Psychological Counselling and Youth and adults and she obtained her Masters in Clinical Psychology by the University of Minho. She is a member of the Portuguese Order of Psychologists and has the title of Specialist in Clinical Psychology with Advanced Speciality in Community Psychology.

Her main focuses of research and practice include the Family and their Communities and the processes underlying flourishing and well-being, positive change, and resilience, particularly in situations of multiple challenges. She explores processes associated with human complexity and positive development and change. In this context, she investigates, from a complex systems approach, special emergent properties of interpersonal relationships and their effects as critical processes associated with human flourishing, change, adaptation, and resilience.

She is interested in the theme of Complexity and the study of Complex Systems, in particular processes of change in social human systems. Her research interests encompass the development of evaluation of theoretical frameworks, as well as pragmatic resources and strategies to support the practice of Complex Thinking applied to the management of change in social systems. She has developed a practice of collaborative action-research and applied research focused on family and community development and in the development of models, resources, and tools for the evaluation and promotion of the potential for change (flourishing and well-being, positive change, and resilience) and of positive parenthood in families, in general, and in multichallenged families with at-risk children, in particular.

She is the author of several family/parenting prevention programs such as "Searching Family Treasure" and "Travelling through lands of parenthood(s)" and of integrative approaches to support multichallenged families, such as the "Integrated Family Assessment and Intervention Model".

Her research is not only of an inter/transdisciplinary nature of her research as it also focused on the processes and practices associated with a positive Inter/Trandisciplinarity, adopting a relational and process-focused approach, informed by a complex systems approach.

She is interested in themes related to Interdisciplinary Methodologies and Methods and to the Philosophy of Sciences. She is an elected member of the Council of the Complex Systems Society (2019–2022) and Associate of the York Cross-Disciplinary Centre for Systems Analysis, of the University of York, UK, where she was a visiting academic (2016–2018). She was a visiting academic in the Department of Health Sciences (2017–2018), of the University of York, United Kingdom.

List of Figures

Chapter 1
Introduction: Thinking Complexity

Abstract The history of Humanity is a history of transformation of the ways humans have coupled with, and acted, in relation to the world (Baggini 2018), in its different dimensions. Different modes of thinking, framed by different value systems, have prompted and promoted distinct modes of action. The necessity to think complexity is widely expressed in the literature (Gershenson and Heylighen 2004). It is visible in the circulation of concepts such as 'complex thinking', 'complexity thinking', 'complex systems thinking', amongst others. However, the definition of these concepts is not always clear and their usage may relate to very different ontological, epistemological and pragmatic stances (Melo et al. 2019). Following the seminal contributions of Morin (1990, 2005, 2014) and his focus on the epistemic implications of Complexity, we defend the development of the concept of Complex Thinking as a particular mode of coupling with the world which is congruent with its complexity. We introduce key features to which, we believe, any proposal for Complex Thinking must attend to, and a number of dimensions entering into its conceptualisation. We introduce our proposal of a framework for complex thinking grounded in a relational perspective and in the following features: differentiation, integration/interconnectedness, recursiveness and emergence.

Keywords Thinking complexity · Complex thinking · Complexity thinking · Complex systems

1.1 Modes of Thinking

The history of Humanity is a history of transformation of the ways humans have coupled with, and acted, in relation to the world (Baggini 2018), in its different dimensions. Different modes of thinking, framed by different value systems, have prompted and promoted distinct modes of action. The world has undergone significant transformations that challenge our capacity for understanding and managing its changes.

Society in general, and Science in particular, in a close relationship of mutual influence (Nowotny, Scott and Gibbons 2001), have come to attribute special value to a restricted set of modes thinking upon which rest the dominant ways of acting in relation to the world, to the neglect of others. Science, in particular, has locked itself in a reduced number of modes of thinking (Chalmers 1999) that, despite great achievements in particular domains, have failed to support the development of a positive co-evolving relationship between humans and their environments and the social worlds they built together. The signs of the lack of congruence in this relationship are increasingly visible as many systems collapse and the challenges to keep positive and sustainable lives increase (Sterman 2006). The limited scope of traditional modes of thinking is more apparent in relation to our attempts to predict and influence systems that, by showing a particular set of properties have been labelled as "complex" (Érdi 2007; Nicolis and Nicolis 2007).

The pressure of finding solutions to the world's most demanding challenges has propelled the development of a diversity of models and simulations and accelerated theoretical, technological and methodological developments (Edmonds and Meyer 2013; Magnani and Bertolotti 2017; Siegfried 2014) aimed at grasping complexity.

Our current capacity to see extraordinary expressions of the complexity of the world, beyond the limits afforded by mechanistic worldviews and reductionist modes of thinking (Capra 1996), has confronted us with a wider world of possibilities for understanding and action. Increasingly, the world is described through lenses that reveal it as changing, dynamic, changing, interconnected, multi-leveled, emergent and largely self-organised (Capra and Luisi 2014). The rise of Systems Thinking, Ecology and of the Complexity Sciences has left a significant mark in our landscape of possibilities for thinking and acting in the world (Capra 1996; Reynolds 2011; Urry 2005). They have highlighted the limitations of the Newtonian worldview and of reductionist approaches (Capra 1996) and invited the development of new modes of thinking that attend to its complexity (Gershenson and Heylighen 2004; Waddington 1977; Morin 2005). But this movement is not without its pitfalls. The values, aspirations and modes of thinking underlying the traditional reductionist and mechanistic paradigms (Capra 1996) have, to some extent, been transported into the realm of the 'Complexity Sciences', under the umbrella of what Morin has called restricted complexity (Morin 2007). Aspirations of control and prediction and the establishment of "Grand Laws" characterise many attempts to approach complexity by finding ways to 'simplify' it (Ib.). The modes of thinking may have become more attuned to seeing some 'complex properties' of the world, but they are not, in themselves, necessarily more complex or congruent with such complexity (Morin 2007; Caves and Melo 2018). The popularised search for impressive models and the misguided belief that more complexity can be addressed with more 'Big' data (Uprichard 2014) has constrained the development of a science of complexity aimed at truly understanding, enacting and, consequently, managing complexity. While the more recent metaphors of the world (e.g. systems; networks; butterfly effect) are significantly different the clockwork images of the Newtonian paradigm (Capra 1996; Capra and Luisi 2014), the dominant ways of exploring and adventuring into those landscapes do not necessarily share the same complex features as those attributed to it. If science continues to unfold in a "restricted mode", the new 'normal science' (Kuhn 1970) of complexity may become trapped in modes of thinking that are inappropriate or inadequate for developing effective means to managing the complexity of natural and social systems towards positive changes (Edmonds 2018; Caves and Melo 2018). Notwithstanding our efforts, the complexity of the world still challenges our understanding and makes change difficult to manage.

1.2 Thinking Complexity

The necessity to think complexity has been expressed in the literature (Gershenson and Heylighen 2004). It is visible in the circulation of concepts such as 'complex thinking', 'complexity thinking', 'complex systems thinking', amongst others. However, as a recent study suggests, the definition of these concepts is not always clear and their usage may relate to very different ontological, epistemological and pragmatic stances (Melo et al. 2019).

Morin has called attention to how some approaches are "restricted" in the way they attempt to approach complexity with the goal of simplifying it (Morin 2007), thereby neglecting the wider epistemological implications of the very notion of complexity and a focus on relations and organisation.

There is a lack of consensus regarding the definition of complexity and there are multiple definitions and perspectives (Manson 2001; Mitchel 2009). We choose to ground the concept of complexity in a relational perspective, connecting it to the following key features: differentiation, integration/interconnectedness, recursiveness and emergence. The world unfolds in a dynamic balance towards progressive differentiation and integration (Heylighen 2008; McShea and Brandon 2010). It organises itself in a multiplicity of recursive, self-referential loops (Clarke 2009; Kauffman 1987; Varela 1984) and into a dynamic network of relations out of which emergence of new properties, patterns, functions and structures emerge (Goldstein 1999) giving rise to new levels of organisation of reality that further make contributions to its differentiation and integration/interconnectedness.

A relational organisation of the world supports its (adaptive) capacity for emergence, through which it generates the kind of novelty (e.g. structures, functions, patterns, properties and dynamics) that cannot easily be reduced to the level of its individual components (Corning 2002; Goldstein 1999) yet influences and constrains them in a recursive manner. As we, as human observers, become more capable of performing a multiplicity of distinctions in the world, and of integrating them, we are also called to develop our of thinking further and to embrace different types of logic. Our challenge is to revise our modes of description and explanation towards greater congruence with the complex properties we are now able to identify in the world (Waddington 1977; Morin 2005). They should be capable of evolving towards increasing differentiation and integration, in a self(-eco-)organised way (Morin 2005), through processes of recursion associated with emergence, opening up new possibilities for action and (self)transformation in the world. As observers (Varela 1976; von Foerster 2003a), and part of the fabric of the world, we co-evolve with it, engaged in structural coupling and mutual determination (Maturana and Varela 1992).

Morin had a seminal contribution to thinking complexity. He advanced with the notion of complex thinking as a mode of thinking that is organised congruently with properties attributed to the complex world (e.g. dialogical principle; hologrammatic principle; recursiveness principle). He has highlighted the epistemic implications of the ideas around complexity for the ways we organise knowledge (Morin 1990; 2005; 2014) and advocated for a "general" approach to complexity to embrace a focus on the relational organisation of systems and our knowledge of them (Morin 2007). Understanding complexity would, thus, require embedding our own patterns of thinking in its apparent paradoxes and loops, integrating uncertainty and departing from the dualistic and reductionist frameworks that fail to acknowledge the complementary nature (of how we construct) reality and the dialectic and dialogical relationship of its complementary pairs (e.g. object ~ subject; part ~ whole; quantitative ~ qualitative; change ~ stability; physical ~ mental[1]) (Bateson 1979; Kelso and Engstrom 2006; Morin 1990), alongside the processes that give rise to them (Varela 1976). Complex thinking would then imply practices of relating to the world that afford both a differentiated and integrated view of reality and the ways we construct knowledge in relation to it. To embrace complexity, the thinking needs to recognise and move across different modes and levels, experiment with multiple ways of making distinction and setting boundaries, aware that setting boundary is associated with the enactment of values and creating constraints that set up a stage for different possibilities for action that bring forth different types of world (Ulrich and Reynolds 2010).

The recognition of the role of the observer in the construction of the complexity of the real (Morin 2005) has long been a basic corolatte of second-order cybernetics (Von Foerster 2003b). This legacy is alive within the tradition of second-order science (Lissack 2017; Müller 2016; Umpleby 2010). Thinking about complexity calls for thinking about the role of the observer in the context of a complementary pair: thinking the (complexity of the) observer(s) ~ thinking the (complexity of) observed realities/world. This could be better presented as a complementary hologramatic pair organised recursively in a relationship of complex causality: thinking the complexity of the observer ~ that thinks/contributes to the complexity of the world ~ that contains/creates the observer that thinks it/creates it.

To think about complexity can be thought of as a form of second order complexity (Tsoukas and Hatch 2001). It is fundamental that our modes of thinking embrace the complementary nature of the world and that we are able to couple with it with sufficient reflexivity to understand our own role in the making and transformation of the same phenomena that we wish to understand, along with the processes that allow for our co-emergence and co-evolution. Understanding the role of the observer when speaking about complexity needs to be tied to the ontological and epistemological foundations for the concept of complexity and of complex thinking. It is necessary to operate with the triad observer-complexity-complex thinking in mind in attending to the processes that support complexity. In some domains of science, particularly within

---

[1]The '~' symbol is used following the proposal of Kelso and Engstrom (2006) to indicate a complementary pair, where both sides are understood only in the context of the relation to each other.

the social sciences, a diversity of practices have been developed that are more attuned to the rich fabric of reality and to the multiplicity of processes and dimensions that sustain it, particularly in relation to its social dimensions and the role of the one that knows in the production of knowledge (Denzin and Lincoln 2017). However, outside the realm of postmodern, constructive, critical and qualitative traditions, the consideration of these dimensions and the role of the observer are (sometimes) added to, but not always fully embraced by mainstream science. Nevertheless, there is a diversity of alternative discourses punctuating the dominant modes of thinking where the traditional modes of science are challenged: to embrace and connect with other modes of knowledge production (Gibbons et al. 1994; Santos 2018), to produce more contextualised and localised knowledge (Lacey 2014) to accept risk and uncertainty (Funtowicz and Ravetz 1994), to coordinate multiple perspectives on reality (Checkland and Scholes 1990; Reynolds 2011), and to consider issues of value, perspective, and power in knowledge production (Muller, 2016; Umpleby 2010; Ulrich and Reynolds 2010). The coordination of alternative modes of knowing, with distinct implications for action, by putting forth a multiplicity of perspectives on the world, may support more complex understandings. Notwithstanding, their coordination and integration in practice might be constrained when the different contributions have emerged from different domains and traditions with distinct ontological, epistemological and pragmatic assumptions. Hence, the multidimensional complex nature of the (complex) world calls not just for different approaches but also for meta-frameworks capable of coordinating different contributions and underlying worldviews, with a focus on practice. A pragmatic[2] focus may be necessary whereby we evaluate the value and "truthfulness" of different modes of thinking through the possibilities for action that they afford as well as their effects and consequences (James 1955). We believe it is necessary to meet complexity with complexity, building modes of coupling that generate progressively more differentiated and integrated constructions and that evolve through a recursive organisation through the integration of their pragmatic effects and the novelty they generate through action.[3]

Emergence (Corning 2002), as a key feature of complexity, is associated with the capacity of many systems to surprise their observers, to change their behaviour, to adapt and evolve (McDaniel and Driebe 2005). It is expressed in the appearance of novelty that calls for new modes of description or explanation as it cannot be reduced to the levels of the behaviour of the parts of a system, albeit constraining them. The emergent novelty is associated with a new organisation of relations where the parts and the whole of a system become simultaneously more and less than each other (Morin 1992).

The concept of emergence, at the core of the notion of complexity, resonates with the creative mind (Boden 2004) and the processes implicated in abductive thinking (Fann 1970; Nubiola 2005) and creative imaginative leaps (Whitehead 1978). Major scientific advances have been propelled by abductive thinking (Rozenboom 1997)

---

[2] "Ideas (which themselves are but parts of our experience) become true just in so far as they help us to get into satisfactory relation with other parts of our experience" (James 1955, p. 49).

[3] Ideas "upon which we can ride" (James 1955, p. 49).

as a distinctive, but complementary, mode of thinking to induction and deduction. In fact, the role of these modes of thinking in scientific development are better understood in the context of recursive cycles of cooperation between them (Gubrium and Holstein 2014; Reichertz 2014).

In many natural, physical and social systems, emergence can be seen as an expression of their complexity. These, so called, "complex systems" exhibit a set of properties (e.g. recursiveness, coordination, non-linearity, robustness) that are critical in sustaining their adaptive capacity to change and evolve in congruence with their environment (Cilliers 1998; Byrne and Callaghan 2014; Érdi 2007; Kelso 1995; Manson 2001).

The study of these properties may lead to a better understanding of the creative and transformative powers of the world, at the core of its complexity, as well as of those of the creative and abductive mind (Boden 2004; Nubiola 2005). At the same time, this understanding is likely to be facilitated by modes of thinking that are somehow congruent with their own organisational principles.

1.3 Thinking Complexity Complexly

We propose that it is possible to organise our modes of thinking in such a way that the resulting coherence fosters the emergence of relevant information about a target system of interest (Checkland 1999; Caves and Melo 2018), ourselves and our relationship with it, so as to improve and expand our possibilities for acting in positive ways in relation to it and to better manage the change processes implicated in this relationship. We expect that more complex modes of thinking are more likely to lead to the emergence of novel and pragmatically[4] meaningful information about a given target system that informs actions supporting a positive co-evolving relationship and eco-systemic fitness between the observer, the system of interest and their environment. The complexity of these modes of thinking may be grounded in properties similar to those attributed to more complex systems. We conceptualise complex thinking as a mode of coupling with the world that attends to its complexity as much as it performs complexity through the enactment of key processes that have been identified in the complex world (as we, as human observers, are currently capable of describing it).

The coordinated exploration of the complexity of the world and the complexity of the thinking may support our growing understanding and capacity to act in relation to both.

It seems necessary to understand the properties of the coupling process and the conditions under which the coupling of an observer with a system of interest generate

---

[4]We broadly adopt a pragmatic view of knowledge and the practice thinking in line with the tradition established by the American Pragmatism, namely the work of Charles Sanders Peirce and William James (cf. James 1955).

meaningful and pragmatically relevant novel information for the management of change of such systems (Rogers et al. 2013; Caves and Melo 2018).

A given observer needs to be sufficiently complex (to practice highly differentiated, integrated and abductive-emergent type of thinking) to grasp a certain expression of the complexity of the world. Following Ross Ashby's Law of Requisite Variety (1957, 1958) one could state that the *more complex the observer/intervenor, in relation to the complexity of a target system/world, the more likely the intervenor will be capable, through their contributions to the coupling relationship, both with the target system and its environments* (Caves and Melo 2018), *of promoting or managing change associated with positive outcomes and positive co-evolution, as identified from the position of a wide number of critical observers.*

The complexity of the observer will bring forth a world that will invite a particular type of practice that, when entering a wider ecology of actions will interact with them in unpredictable ways (Morin 1992). By approximating complexity with complexity, the actions may be more capable of fitting and evolving with this wider ecology and the thinking may generate a sense of coherence, that may be thought of as a type of "glue" that holds complex systems together and that, in relation to our explanations, generates simplicity out of the complexity of our relation with them (Lissack and Graber 2014).

The study of complexity and the study and practice of complex thinking can support and propel the development of each other. Complexity scientists and philosophers have already opened a space of dialogue regarding the philosophical implications of complexity (Gershenson and Heylighen 2004; Heylighen, Cilliers and Gershenson 2006; Morin 2001). This space needs to be further expanded and explored in ways that allow for a refinement of the notion of complex thinking towards its pragmatic development. This movement is fundamental for the operational conceptualisation of complex thinking towards providing guidance into practice for the development of interventions, strategies and resources to promote and evaluate it.

In this context, the operational definition of the properties of complex thinking requires a clarification of what 'more complexity' might mean in this context. As for other types of 'complexity' it seems relevant to promote further discussion about the ontological and epistemological assumptions grounding the use of the concept (Richardson 2005; Lissack 2014; McIntyre 1998) particularly considering its implications for action. This discussion about complex thinking necessary refers back to the notion of complexity per se, and the debate of what 'complexity' refers to: is it a 'real' property of the world, of the observer, or both, or is it something else? (Casti and Karlqvist 1986; Midgley 2008). A set of foundational questions need to be addressed to further clarify the notion of complex thinking in relation to its pragmatic implications. In particular, it is important to address questions of an ontological nature, such as 'What is the nature of the complexity of the thinking and where can it be found?' and epistemological interrogations such as 'how do we identify/know about/measure the complexity of the thinking?'.

It seems necessary to further deepen the theoretical foundations of the notion of complex thinking in ways that (i) supports (a set of) definition(s) that both distinguish and relate complex thinking to similar concepts used in the literature, including the notion of complexity itself; (ii) establishes an ontological and epistemological position that empowers the heuristic value of the concept of complex thinking and affords novel possibilities for action in relation to change in complex systems; (iii) informs a pragmatically relevant operationalisation of the concept in ways that may inform practices to support it, evaluate and investigate it further.

We aim to contribute to address these objectives by conceptualising complex thinking as a mode and outcome of coupling with the world that both attends to and enacts complexity. This notion of complex thinking is grounded in a relational and constructivist worldview capable of bridging and coordinating (distinguishing and relating) different pragmatic contributions for the operationalisation and practice of complex thinking stemming from distinct ontological and epistemological traditions.

References

W.R. Ashby, *An Introduction to Cybernetics* (Chapman & Hall Ltd, London, 1957)

W.R. Ashby, Requisite variety and its implications for the control of complex systems. Cybernetica **1**, 83–99 (1958)

J. Baggini, *How the World Thinks. A Global History of Philosophy* (Granta, London, 2018)

G. Bateson, *Mind and Nature: A Necessary Unity* (Bantam Books, New York, 1979)

M.A. Boden, *The Creative Mind: Myths and Mechanisms* (Routledge, London, 2004)

D. Byrne, G. Callaghan, *Complexity Theory and the Social Sciences: The State of the Art* (Routledge, London, 2014)

F. Capra, *The Web of Life: A New Scientific Understanding of Living Systems* (Anchor, 1996)

F. Capra, P.L. Luisi, *The Systems View of Life: A Unifying Vision* (Cambridge University Press, 2014)

J.L. Casti, A. Karlqvist, Introduction, in *Complexity, Language and Life: Mathematical Approaches* (xi-xiii), eds. by J.L Casti, A. Karlqvist (Springer-Verlag, Berlin, 1986)

L. Caves, A.T Melo, (Gardening) Gardening: A relational framework for complex thinking about complex systems, in *Narrating Complexity*, eds. by R. Walsh, S. Stepney (Springer, London, 2018), pp. 149–196. https://doi.org/10.1007/978-3-319-64714-2_13

P. Cilliers, *Complexity and Postmodernism. Understanding Complex Systems* (Routledge, London, 1998)

A.F. Chalmers, *What is this Thing Called Science?*, 3rd edn. (Hackett Publishing Company Inc., Queensland, 1999)

P. Checkland, *Systems Thinking, Systems Practice* (John Wiley, Chichester, 1999)

P. Checkland, J. Scholes, *Soft Systems Methodology in Action. Includes a 30-year Retrospective* (John Wiley & Sons, Chichster, 1990)

B. Clarke, Heinz von Foerster's Demons. The emergence of second-order systems theory. in *Emergence and Embodiment*, eds. by B. Clarke, M. Hansen (Duke University Press, Durham, NC, 2009), pp. 34–61

P.A. Corning, The re-emergence of "emergence": a venerable concept in search of a theory. Complexity **7**(6), 18–30 (2002)

N.K. Denzin, Y.S. Lincoln, *The SAGE Handbook of Qualitative Research* (SAGE Publications, 2017)

B. Edmonds, System farming, in *Social Systems Engineering: The Design of Complexity*, ed. by C. García-Díaz, C. Olaya (John Wiley & Sons, Chichester, 2018), pp. 45–64

B. Edmonds, R. Meyer, *Simulating Social Complexity* (Springer-Verlag, Berlin, 2013). https://doi.org/10.1007/978-3-540-93813-2

P. Érdi, *Complexity Explained* (Springer Science & Business Media, 2007)

K.T. Fann, *Peirce's Theory of Abduction* (Martinus Nijhoof, The Hague, 1970)

S.O. Funtowicz, J.R. Ravetz, Uncertainty, complexity and post-normal science. Environmen. Toxicol. Chem. SETAC **13**(12), 1881–1885 (1994)

C. Gershenson, F. Heylighen, *How can we think the complex? arXiv [nlin.AO]* (2004). Retrieved from http://arxiv.org/abs/nlin/0402023

M. Gibbons, C. Limoges, H. Nowotny, S. Schwartzman, P. Scott, M. Trow, *The New Production of Knowledge: The Dynamics of Science and Research in Contemporary Societies* (Sage, London, 1994)

J. Goldstein, Emergence as a construct: history and issues. Emergence **1**(1), 49–72 (1999)

J.F. Gubrium, J.A. Holstein, Analytic inspiration in ethnographic fieldwork, in *The SAGE Handbook of Qualitative Data Analysis*, ed. by U. Flick (Sage, London, 2014), pp. 35–48

F. Heylighen, *Complexity and Evolution. Fundamental concepts of a new scientific worldview.* Lectures notes 2017–2018 (2018). Retrieved from http://pespmc1.vub.ac.be/books/Complexity-Evolution.pdf

F. Heylighen, P. Cilliers, C. Gershenson, *Complexity and Philosophy. Complexity, Science, and Society* (2006). http://cogprints.org/4847/

W. James, *Pragmatism and Four Essays from the Meaning of Truth* (New American Library, New York, 1955). [Pragmatism originally published in 1907; The Meaning of Truth originally published in 1909]

L. Kauffman, Self-reference and recursive forms. J. Soc. Biol. Struct. **10**, 53–72 (1987)

S.J.A. Kelso, *Dynamic Patterns: The Self-Organization of Brain and Behavior* (MIT Press, Cambridge, MA, 1995)

S.J.A. Kelso, D.A. Engstrom, *The Complementary Nature* (MIT Press, 2006)

T.S. Kuhn, *The Structure of Scientific Revolutions (2nd Edition, Enlarged)* (The University of Chicago Press, Chicago, 1970)

H. Lacey, Scientific research, technological innovation and the agenda of social justice, democratic participation and sustainability. Sci. Stud. **12**(SPE), 37–55 (2014)

M. Lissack, The context of our query. in *Modes of Explanation. Affordances for Action and Prediction*, ed. by M. Lissack, A. Graber (Palgrave Macmillan, New York, 2014), pp. 25–55

M. Lissack, Second order science: examining hidden presuppositions in the practice of science. Found. Sci. **22**(3), 557–573 (2017)

M. Lissack, A. Graber, in *Preface*, eds. by M. Lissack, A. Graber. Modes of Explanation. Affordances for Action and Prediction (Palgrave Macmillan, New York, 2014), pp. xviii–xvi

R.R. McDaniel, D. Driebe, *Uncertainty and Surprise in Complex Systems: Questions on Working with the Unexpected* (Springer, Berlin, 2005)

L. McIntyre, Complexity: a philosopher's reflections. Complexity **3**(6), 26–32 (1998)

D.W. McShea, R.N. Brandon, *Biology's First Law: The Tendency for Diversity and Complexity to Increase in Evolutionary Systems* (University of Chicago Press, 2010)

L. Magnani, T. Bertolotti, *Springer Handbook of Model-Based Science* (Springer, Switzerland, 2017). https://doi.org/10.1007/978-3-319-30526-4

S.M. Manson, Simplifying complexity: a review of complexity theory. Geoforum J. Phys. Hum. Reg. Geosci. **32**(3), 405–414 (2001)

H. Maturana, F. Varela, *The Tree of Knowledge. The Biological Roots of Human Understanding* (Shambhala, Boston, MA, 1992)

A.T. Melo, L.S,D. Caves, A. Dewitt, E. Clutton, R. Macpherson, P. Garnett, Thinking (in) complexity: (In)definitions and (mis)conceptions. Syst. Res. Behav. Sci. **37** (1), 154–169, (2019). Doi:doi.org/10.1002/sres.2612

G. Midgley, Systems thinking, complexity and the philosophy of science. Emerg. Complex. Organ. **10**(4), 55–73 (2008)

M. Mitchell, *Complexity: A Guided Tour* (Oxford University Press, 2009)

E. Morin, *Science Avec Conscience. Nouvelle edition* (Fayard, Paris, 1990)

E. Morin, From the concept of system to the paradigm of complexity. J. Soc. Evolut. Syst. **15**(4), 371–385 (1992)

E. Morin, in *The Epistemology of Complexity*. eds. by F.D. Schnitman, J. Schnitman. New Paradigms, Culture and Subjectivity (Hampton Press Inc, New York, 2001), pp. 325–340

E. Morin, *Introduction à la pensée complexe* (Éditions du Seuil, Paris, 2005). [originally published in 1990]

E. Morin, in *Restricted Complexity, General Complexity*. ed. by E. Gersherson, D. Aerts, B. Edmonds. Worldviews, Science and Us. Philosophy and Complexity. (World Scientific, London, 2007), pp. 5–29

E. Morin, Complex thinking for a complex world–about reductionism, disjunction and systemism. Systema Connect. Matter Life Cult. Technol. **2**(1), 14–22 (2014)

K.H. Müller, *Second-order Science: The Revolution of Scientific Structures*. (Echoraum Wien, 2016)

G. Nicolis, C. Rouvas-Nicolis, Complex systems. Scholarpedia, **2**(11):1473 (2007). doi:10.4249/scholarpedia.1473

H. Nowotny, P. Scott, M. Gibbons, *Re-thinking Science. Knowledge and the Public in the Age of Uncertainty* (Blackwell publishers, Cambridge, 2001)

J. Nubiola, Abduction or the logic of surprise. Semiotica **153**(1/4), 117–130 (2005)

J. Reichertz, Induction, deduction, abduction, in *The Sage Handbook of Qualitative Data Analysis*, ed. by U. Flick (Sage, London, 2014), pp. 121–135

M. Reynolds, Critical thinking and systems thinking: towards a critical literacy for systems thinking in practice, in *Critical Thinking*, ed. by C.P. Horvath, J.M. Forte (Nova Science Publishers, New York, 2011), pp. 37–68

K. Richardson, The hegemony of the physical sciences: an exploration in complexity thinking. Futures **37**(7), 615–653 (2005)

K.H. Rogers, R. Luton, H. Biggs, R. Biggs, S. Blignaut, A.G. Choles, C.G. Palmer, P. Tangwe, Fostering complexity thinking in action research for change in social–ecological systems. Ecol. Soc. **18**(2), 31 (2013). https://doi.org/10.5751/ES-05330-180231

W.W. Rozenboom, in *Good Science is Abductive not Hypothetical-Deductive*, eds. by L.L. Harlow, S.A. Mulaik, J.H. Steiger. What if there were no Significance Tests?. (Erlbaum, New Jersey, 1997), pp. 366–391

B.S. Santos, *The End of the Cognitive Empire: The Coming of Age of Epistemologies of the South* (Duke University Press, 2018)

R. Siegfried, *Modeling and Simulation of Complex Systems: A Framework for Efficient Agent-Based Modeling and Simulation*. (Springer, 2014)

J.D. Sterman, Learning from evidence in a complex world. Am. J. Public Health **96**(3), 505–514 (2006)

H. Tsoukas, M.J. Hatch, Complex thinking, complex practice: the case for a narrative approach to organizational complexity. Hum. Relat. Stud. Towards Integr. Soc. Sci. **54**(8), 979–1013 (2001)

W. Ulrich, M. Reynolds, Critical Systems Heuristics, in *Systems Approaches to Managing Change: A Practical Guide*, ed. by M. Reynolds, S. Holwell (Springer, London, London, 2010), pp. 243–292

S. Umpleby, From complexity to reflexivity: underlying logics used in science. J. Wash. Acad. Sci. **96**(1), 15–26 (2010)

E. Uprichard, Big Doubts about Big Data, The Chronicle of Higher Education (2014), October 2013

J. Urry, The Complexity Turn. Theory, Culture & Society **22**(5), 1–14 (2005)

F. Varela, *Not one, not two* (The Coevolution Quarterly, Fall, 1976), pp. 62–67

F. Varela, The creative circle: sketches on the natural history of circularity, in *The Invented Reality*, ed. by P. Watzlawick (Norton Publishing, New York, 1984)

Von Foerster (2003a). Notes on an epistemology for living things [Address originally given in 1972]. In Von Foerster, H. (2003). *Understanding Understanding: Essays on Cybernetics and Cognition* (Springer-Verlag, New York, 2013), pp. 247–259

Von Foerster (2003b). On constructing a reality [Address originally published in 1973]. In Von Foerster, H. (2003). *Understanding Understanding: Essays on Cybernetics and Cognition* (Springer-Verlag, New York, 2003), pp. 211–227

C.H. Waddington, *Tools for Thought* (Paladin, St. Albans, 1977)
A.N. Whitehead, *Process and Reality*, corrected edn. (The Free Press, New York, 1978)

Chapter 2
A Relational Framework for Complexity and Complex Thinking

Abstract In this chapter, we make a brief presentation of the ontological and epistemological assumptions of a relational worldview that provide the grounds for a relational conceptualisation of complexity and complex thinking and for bridging and integrating different perspectives. We discuss the role of the observer in the construction of Complexity and its implications for how to define and think about the complex. We highlight the role of the nature of the coupling between the Observer and the world in producing knowledge and guiding the action of the Observer. From a relational perspective on complexity, we proceed with a discussion of the relative status of statements about complexity and the role of numerous constraints in the construction of knowledge, including the observer's own structural determination. We discuss the implications of a relational worldview for making judgments about complexity, namely the complexity of the thinking.

Keywords Relational worldview · Observer · Complexity · Complex thinking · Relativity

2.1 What and Where Is Complexity?

Although debates around complexity raise important philosophical and pragmatic issues (Heylighen et al. 2006), there is not always an explicit reference to the ontological and epistemological positions framing notions of complexity and the meaning of complex thinking. Rosen has defined a complex system as that for which multiple modes of description and multiple perspectives need to be adopted, and no single model suffices (Rosen 1977, 2000). We believe the notion of complexity, itself, requires the integration of a variety of ontological and epistemological perspectives.

In understanding reality, there is a general divide between a broad category of realists, on one hand, and constructivists, on the other (Richardson 2005; Lissack 2014). For the former, complexity could be as seen an inherent, 'real' property of the systemic entities that compose the universe, as proposed by complex realism (Williams and Dyer 2017; Byrne and Callaghan 2014). Many complexity theorists assume that the complex systems are 'real entities' that compose the world as we know it (Byrne and Callaghan 2014). For the latter, complexity is located on the

A. Teixeira de Melo, *Performing Complexity: Building Foundations for the Practice of Complex Thinking*, SpringerBriefs in Complexity,
https://doi.org/10.1007/978-3-030-46245-1_2

side of the observer, belonging to the realm of their constructions (Lissack 2014). Systems are then seen as a way of thinking and organising information about the world (Reynolds 2011). For others, like Morin, one cannot dissociate the world from its observers, and systems as real entities from the minds that identify or construct them (Morin 1992).

These positions are also related to the question of where "to find" complexity. Some would position complexity as an epistemological issue arguing that the world is not, in itself complex (it 'just is'), but only "as described and defined by a given level of inquiry" (McIntyre 1998, p. 28). Others come to see complexity as defined at the level of the relation between an observer and the world and expressed in the difficulties associated with modelling and prediction (Edmonds 2000) in systems that afford multiple modes of interactions with the observer (Rosen 1977). Complexity can then be expressed as a relational property: "it is probably more meaningful to consider complexity as a property of interaction than of the systems, although it is clearly associated to them" (Casti 1986, p. 147). When considered as a relational property, complexity is "latent or implicate property of a system" a property "made manifest by the interaction of the [given] system with another" (Casti 1986, p. 146).

We believe it is possible to coordinate and articulate different views. Richardson (2005) has attempted to build a pluralistic framework from a realist ground. Others have proposed indicators of complexity based on a relational ontology (Heylighen 2018). We attempt to build an integrative position adopting a relational constructivist worldview that, we believe, may be capable of bridging approaches stemming from realism and constructivism arguments at an ontological and epistemological level. From this perspective we might be able to accept that there is a 'real world' 'out there' and that it is (to some extent), independent of (at least some) observers. On the other hand, from this perspective it is also necessary to assume that even though it may not be dependent on particular observers, nothing exists or can be known independently of something else (Whitehead 1978). We assume the world exists as a vast relational matrix where every entity is simultaneously constituted and sustained by relationships with other entities to which it contributes in the context of relations that, recursively, also given rise to them as distinct entities. Much of the causal thinking engaged in complexity requires a recognition of the critical role of recursiveness and how it challenges conventional logic. Congruently, it is necessary to accept recursiveness as a fundamental dimension both of the organisation of the world and our knowledge of it.

In an infinite network of relations the influence of some relations and entities over others will be largely dependent on their relative position. The role of the observers as special entities in such a network is particularly relevant as they create the textures of reality by extracting patterns and marking distinctions, punctuations and indications that shape the relational processes that bring forth a particular reality for themselves and others, which offer particular possibilities for action (Goguen and Varela 1979). The observers have emerged out the relational matrix and the processes that organise the universe (Whitehead 1978). The observers' acts are not just reflections of the 'real' nature of this universe, as this one affords many possibilities, including their existence. Neither are they fully independent of it, as they belong to the vast network of processes from which they extract, as patterns, the 'things' that compose their world. Even though we assume that (a large part of) the world may be considered to exist independently of a (set of particular) observer(s), we also assume it does not exist outside of a network of relations where each entity is somehow constituted and dependent on some other entity. The observers, in turn, through their actions in this network, are responsible to shape, arrange or transform (at least) some of the configurations of relations through which the world appears to them in a given way and, by coordination with others, they provide stability to the boundaries that sustain them (Richardson 2005). We will further explore these ideas in relation to a relational understanding of complexity and our proposal for complex thinking.

2.2 A Relational Worldview for Complexity: From Differentiation and Integration, to Recursiveness and Emergence

Our conceptions of complexity and complex thinking are based in a relational, constructive and processual worldview, where nothing can be thought to exist or to be known outside of a relation (Bateson 1979; Whitehead 1978) and where "things" need to be understood as punctuations in a complex stream of dynamic relational processes.

Albeit it is not our intention to review the different proposals of relational and process worldviews, it is necessary to note that it has underlied the work of many authors, namely in the understanding of the processes of life, nature and human development. Process-relational philosophy (Mesle 2008) establishes assumptions that can be found in different proposals in many other domains but the development of a relational worldview in these domains has also contributed to enrich and strengthen the philosophical debate on this subject. Rosen, following the work of Rashevksy, has set up the foundations of a Relational Biology making important contributions to the realisation of a relational worldview (Rosen 1991, 2000). Likewise, Developmental Systems theorists (Oyama et al. 2001; Lerner 2002) and many proponents of organismic and contextualist worldviews, can be seen as some of many "faces" of relational approaches (Stetsenko 2017). Stetsenko (2017) calls attention to the common relational grounds of the pragmatist proposal of Dewey and its implications for understanding knowledge and education and the work of developmentalists such as Piaget and Vygostsky, highlighting the role of action and its inherently relational nature.

A relational worldview has also had some expression in debates about complexity (Heylighen 2018). We are interested in exploring a relational worldview as it provides the foundations to ground a conceptualisation of complexity that, in turn, supports our proposal for complex thinking. We believe it has important pragmatic implications for how to manage our relationship with change and that it may bridge the dialogue between different views of complexity. From this position, everything in the world is said to be in a mutually defining relation to something else, and all relations have, at their core, some (relative) degree of difference ('that makes a difference') (Bateson 1979). From this perspective, a relation is considered to be the basic unit for the consideration of complexity, and it is the organisation of these relations (e.g. how relations relate to each other) that lies at the core of complexity (Morin 1992). It is in the context of relations that all things exist and are sustained, through processes of mutual co-arising and co-determination (Macy 1991). Even the origins of the Universe could be hypothesised to be tied to the emergence of (at least) one primordial relation (or set of relations), constituted as differences that generated further differences (differentiated). In this network of relations, processes of recursive cooperation and coordination led to the formation of higher order relational entities (integration) associated with new (emergent) properties.

We can assume that we talk about the complexity of the universe because it has differentiated and integrated and given rise to new entities, as parts and wholes that are both more and less than each other (Morin 1992). The universe then needs to be understood in relational terms. There are a variety of attempted definitions of complexity and classifications ('types') of complexity, as well as debates on how to measure complexity (Ladyman et al. 2013; Manson 2001). We believe that a core definition of complexity can be tied to notions of differentiation, integration (McShea and Brandon 2010; Tononi et al. 1994; Lerner 2002) recursiveness (Kauffman 1987; Von Foerster 2003a) and emergence (Corning 2002).

The different elements in the universe are organised in relational structures that are progressively differentiated, and integrated, through processes of coordination leading to the emergence of new properties that can be distinguished as a new level, or a new type of 'thing', with further horizontal and vertical movements of differentiation. Differentiation and integration can then be seen as fundamental properties associated with many ways relations can be organised, underlying one possible way to conceptualise complexity (Casti and Karlqvist 1986; McShea and Brandon 2010).

The creative power of the universe in visible in the generation of a diversity of relations and entities (differentiation), that coordinate to constitute new wholes (integration), through non-linear couplings and recursive loops (recursiveness), through which new levels of organisation appear with novel and non-reducible properties (emergence). The complexity of the universe is hence revealed in the weaving of a fabric that is as much fluid, dynamic and disordered as it is organised, multilayered and ornamented by the patterns emergent from the creative arrangements and configurations of differences/information/relations that constitute its basic threads.

2.3   The (Co)Evolution of Complexity: The Universe and Its Observers

A relational worldview of the Universe sustains arguments of a complementary nature for its ontology and epistemology (Kelso and Engstrom 2006). Everything that exists, sustained by relational processes, can only be known through complementary processes, that oftentimes appear as dualities, in the acts of distinction made by human observers. These dualities, operating as cybernetics complementaries where one end defines the other, are expressions of underlying processes and distinctions that, operating at other levels are, themselves, sustained by other complementaries (Keeney 1983; Varela 1976). In this entangled and recursive reality, the 'things' that are known can be said to be real but not to be known outside of a relation where that 'which is' is mutually defined by that 'which it is not' (dark–light; night–day; man–women; human–non-human). The choice of the complementary pair defines the identity of the processes.

The things that are perceived as real correspond to snapshots of dynamic processes that are commonly captured and reified in the act of distinction that extracts them, as patterns, from a background of (otherwise perceived) noise. These acts of distinction are performed by observers, themselves an expression of the complexity of the world.

The Universe has also realised its complexity by taking a recursive turn in performing the foundational act of distinction leading to the emergence of an observer. The capacity to identify these properties is itself associated with processes of differentiation, integration, recursiveness and emergence related to an observer.

The observers are, in many ways, a part of the complexity of the universe and of its origins. They set forth the shape of the Universe, through their acts of distinctions and indications (Goguen and Varela 1979) changing their own world and themselves in the process and, in consequence, leading to further changes in the universe of which they are a part (that exists embedded and supported in relations).

Following Maturana's and von Foerster's corollaries we may say: everything that is known, must be known by an observer to an observer (Von Foerster 2003b; Maturana 1978) in the context of a network of relations that configures both. The first recursive loop of the Universe onto itself must have been a fundamental condition for the first distinction to be made and, with it, for an observer to emerge as an entity capable of embodying a particular point of view in that universe (Goguen and Varela 1979). This point of view is relative to a set of embedding relations that create conditions for it to be perceived and sustained as such. In fact, relations sustain different things but 'things' only exist with a particular identity in the eyes of an observer capable of stabilising the acts of distinction and negotiating those boundaries with others in order to create the meanings that transform a relational process into a 'real entity'. These boundaries are distributed and sustained, with different degrees of stability, across communities of observers (Richardson 2005). It is in a dynamic equilibrium that the objects of the Universe are sustained by a network of relations and coupling processes that create particular types of affordances (Gibson 1986) for different communities of observers who enter this vast network and to whose constitution they contribute.

Varela (1976) described at least three defining features of an observer as an entity, presenting: (i) the capacity for indication (to distinguish boundaries, define criteria for boundary stability of an identified system, defining the systems and its modes of operation and attribute values); (ii) the capacity for time (to choose a sequence of segments of time, to chop a net of related events, to compute through a process, to define timescales and approximate the stability of a whole through punctuations in time); (iii) and the capacity for agreement (to externalise views, coordinate and synchronise it with others, to connect and reproduce other/s distinctions and selected time patterns). Others authors have acknowledged the role of the observer in the constitution of a system, which becomes a system-of-interest, through the acts of distinction that define it and the attribution of value that shapes the consequences of such distinctions and the implications for the relationships to be established in it and around it (Checkland and Scholes 1990; Reynolds 2007; Ulrich 2010). The properties listed by Varela are associated with the main feature of an observer which is the capacity for engagement or coupling, as a distinct entity, capable of experiencing itself as an entity, of recognising others as such (through the acts of distinction), and of establishing some degree of interdependence (coordination) with others. This capacity can be described in terms of the properties that characterise the observer's contributions to the coupling and how it is subject to change in the context of a broader environment. It can also be seen both as a process and an outcome that feeds back on itself in a recursive loop. It is in this recursiveness that the coupling generates information that leads to the self-organisation of the entity, and the definition of their boundaries, shaped in the relationship with the other entity.

As the observers act in the Universe and couple with each other, they generate more distinctions and more information that adds to the diversity of the Universe and becomes integrated in its relational matrix. In this recursive dance, information is created, in the form of relations that represent differences that make a difference as they generate perturbations and induce changes. Between movements of differentiation and integration, punctuated with the emergence of novelty, and in the context of a recursive relationship with itself, the universe, the observers and their available modes of coupling are perturbed, change and co-evolve. Figure 2.1 illustrates an abstracted representation of the co-evolution of the complexity of the universe and the observer.

On the left side of the figure one can see the evolution of the Universe in terms of its differentiation, integration and emergence of new dimensions and entities that exist only in relation to each other. On the right side, the figure represents observers and communities of observers, as part of the world, developing an increasing capacity to make distinctions and to recognise the complexity of the universe in their acts of distinctions. Through the observers' capacity for coupling and engagement, these distinctions increase in differentiation and integration, shaping the world and leading to transformations in the relations between the entities that it gives rise to.

These interactions may lead to the emergence of new dimensions in the world, of which they then become a part, as affordances for new distinctions to be drawn and

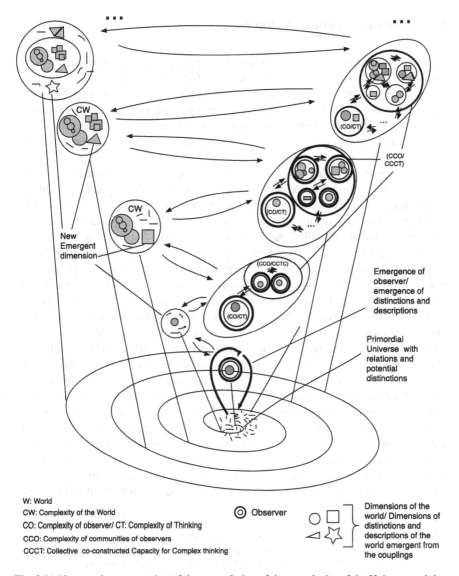

W: World
CW: Complexity of the World
CO: Complexity of observer/ CT: Complexity of Thinking
CCO: Complexity of communities of observers
CCCT: Collective co-constructed Capacity for Complex thinking

◎ Observer

Dimensions of the world/ Dimensions of distinctions and descriptions of the world emergent from the couplings

Fig. 2.1 Abstracted representation of the co-evolution of the complexity of the Universe and the Observers

new relations to be crafted. Hence, the complexity of the world and of the observer co-evolve, in recursive movements, each simultaneously acting as a generative process and a product of the other.

2.4   Knowing Through Coupling

Stetsenko (2017) has called attention to how the proposals of a relational worldview underlying the works of authors such as Vygotsky, Dewey and Piaget, and their common focus on action, introduce a new way of perceiving the observer, beyond a "spectator" role. The author recalls how in these proposals there is "the realization that the only access people have to reality is through active engagement with and participation in it, rather than through simply "being" in the world" (Stetsenko 2017, p. 151).

With the observer's inherent capacity for engagement or coupling comes the potential for change. As both the system of interest and the observer respond to the differences/information generated in the coupling (Maturana and Varela 1992), including the effects of their emergent synergies (Haken 1973), they undergo transformations.

Maturana and Varela (1992) have showcased the idea that living self-organised systems cannot be externally controlled or instructed but only perturbed through coupling with another entity. In the coupling processes all interacting entities are more or less subject to change, depending of the stability ~ instability of the patterns of their interactions and of the properties and organisation of the embedding network of relations that constitutes (through their distinctions) their environment. Some relations will have a more direct or indirect impact, associated with their position in a network of relations, and some will have a more positive contribution (in determining what the other entity is) or a negative one (in determining what the other entity is not or by shaping the environment but not directly affecting its constitution) (Whitehead 1978). These relations will vary in their contributions to the stability or transformation of the observers and their systems-of-interest.[1] Namely their importance to how the system-of-interest is identified and how its boundaries are (re)defined, recognised and sustained by a given community (or communities) of observers depending on the position of the observer-of-interest[2] in that network and on the stability of the relational matrix in which other entities participate in creating affordances for the coupling observer—system-of-interest that allow for particular types of distinctions.

---

[1] The idea of System of Interest stems from the tradition of Systems Thinking, namely Soft Systems, and conveys the notion that a system exists as a set of purposeful activities identified by an observer with a particular interest (Checkland and Scholes 1990; Reynolds 2007). The observer draws certain distinctions and extracts information that allows them to construct the system in a particular way and those distinctions are guided by particular interests and purposes. We adopt this expression to highlight the fact that a system can be considered to have a particular ontological existence, sustained by a particular set of relations but also that this is necessarily related to its existence as an epistemological device and relational object. In this sense, sense, by participating in different ways in those relations that sustain the (ontological) existence of a system (of multiple possibilities), different observers will perform different acts of distinction, that will bring about different types of system. The nature of these distinctions is related to the observer's own relational organisation, preferences, habits, intentions and purposes.

[2] This is defined analogously to the notion of system of interest. It pertains to the observer that, identified (by themselves and/or others) in the context of a particular set of relations and from a particular perspective is of interest, for a particular purpose.

Different modes of coupling will create different types of information, supporting different modes of knowing and different outcomes, leading to different types of transformation (and definition) of the observer, the system and their coupling (Caves and Melo 2018). Hence, while making distinctions and value judgments observers set up an array of possibilities on the basis of which they will interact with other systems of interest and that will determine the type of information created and the outcome of the coupling. By further exploring their capacity for engagement and experimenting with different properties of their contributions for coupling, namely the lenses and perspectives on the basis of which they perform distinctions and indications, the observers will expand or constrain their possibilities for action and change. Systems thinking has long explored this property of observers regarding the possibilities for change that are created by attending to multiple perspectives in the construction of the system-of-interest and to the ways their coordination can open alternative possibilities for action (Reynolds and Holwell 2010).

The effects of the coupling will be dependent on how the structure of the interacting entities determines the information that, generated in that coupling as differences and further affordances for distinctions, is available to them. The degree of congruence in the coupling, and the capacity to make punctuations and attribute meaning to it, is likely to impact its outcomes. Different types of recursive relations can be established in the coupling process that lead to the emergence of information that will influence how the different entities adjust and respond to each other, through their own internal transformations. The coupling will promote changes to the extent that it generates sufficient differences to make a difference in those entities/observer and increases their complexity through greater differentiation and integration of their thinking in relation to the target system. The coupling and the nature of the information produced will subsequently change the observer, the system of interest and their future coupling.

The coupling process may generate such coherence that changes may be minimal or appear so coordinated and synchronised that the behaviour between the interacting entities is better understood through the emergent synergetic properties of their relationship (Kelso 2009). The observer and system may experience a sense of familiarity, of 'knowing' each other, adjusting their responses in the way of 'smooth' movements in a coordinated dance where the movement of one is easily followed the movements of the other in sustaining an overall coherence. They might be so strongly coupled and change so congruently that they develop a capacity to anticipate the response of the other, or at least to adjust to them, so as to preserve the coherence of the whole they now constitute. It may be that coherence is built following the emergence of a perturbation, as a difference that signalled a difference in the coupling process and that produced information that supported learning. In such situations there might be favourable conditions for a type of co-evolution that supports positive change outcomes for the entities involved, as recognised by themselves and other critical observers. Some degree of perturbation may be necessary to generate minimal differences/information that supports continuous coordinated adjustments and the strength and quality of the coupling towards increased coherence.

When there is little congruence in the coupling there might be such an excess of differences/information being generated that it may not be possible to integrate it into the coupling relationship. It is possible that, in some cases, instead of supporting learning and co-evolution, the information created has a destructive role in the organisation of one or both of the entities, and/or in their coupling relationship, impairing the emergence of a higher level of organisation that integrates them in positive ways (Maturana and Varela 1992). The perturbation introduced by the coupling may also become unmanageable, particularly when the observers are less complex than the system of interest, or the coupling is too weak (Caves and Melo 2018; Casti 1986).

The number and nature of the modes of coupling, available to produce information, and their strength, are relevant for the type of outcomes to be achieved. The observer may need to attempt to match the current level of complexity of the target system (differentiation, integration and emergence- cf. ahead) so as to avoid being locked in a state where anticipation and prediction become harder due to the absence of meaningful information. If the target system changes at a higher rate than the observer is capable of adjusting to, the process of change may also be unmanageable. Hence, meaningful differences/information may be created in a close coupling relationship as long as there is some congruence between observer and system-of-interest and with the networks of relationships that embed them as well as the conditions for them to coordinate towards increased coherence. To be an effective agent of change, the observer might need to experiment with diverse modes of coupling in order to produce sufficient variety of information and also closely monitor their own contributions, the responses of the system-of-interest, along with the coupling dynamics and its effects in terms of the transformations experienced in themselves, the target system and in the coupling (Ashby 1958; Caves and Melo 2018).

2.5   From Relational to Relative Complexity(Ies)

From a relational constructive worldview it is possible to articulate a multiplicity of views on complexity. Complexity can be considered as (if it was) a 'real' property of world as much as a property of the observer, but these positions are framed by a wider view where they are necessarily understood as relative to each other or as expressions of a (relational) property of the coupling. From this perspective it is possible, under certain conditions, and for given pragmatic purposes, to refer to these "complexities" independently, as if they were absolute. Complex thinking can thus be treated as if it was an absolute property of the observer. Nevertheless, it will be better understood as a relational property that characterises the relationship between the observer and the world and can be evaluated in the context of a wider network of relations sustaining communities of observers and relationships between systems-of-interest and between observers and systems.

The relations between these 'complexities' (observer, world, relation) underlies the complexity of the universe and its evolution toward increasing differentiation, integration and capacity for emergence. In the wider pool of relations not all are equally relevant or critical in sustaining a given reality as is experienced by particular communities of observers. Let us assume that in the Universe, as a vast, dynamic relational matrix, there are different configurations of relations that more closely sustain a particular entity x as perceived by a particular observer Ox (the observer capable of identifying/relating to x as x) or a community of observers COx (the community of observers sharing the capacity to identify/relate to x as x). Let us also assume that there are diverse observers and communities of observers and other entities that, with different degrees of involvement and different contributions, participate in relations that more directly sustain a section of a wider network of relations that allows for x to be identified with x attributes, and that we may call Netx. It may happen that a particular community of observers A, participating in, or being in a position near that particular region of a wider relational space, is capable of identifying objects or systems-of-interest through a configuration of relations and with properties that a community of observers B cannot grasp but that appear to be partially congruent with those available to the community of observers C. To some extent there is a reality 'out there' for the community B to "discover", but a reality that exists for A and partially for C in a given way and is sustained by NetX. This may be considered as a 'real reality' that exists, at least as a potentiality, and is sustained by Netx and is relatively independent of the communities of observers A, B and C but that comes forth as x in the coupling with community A and partially with community C. Communities A and C would be able to identify that part of reality as something of the nature of x but it will exist in a very different way, or not exist at all for B. It is possible that A and C and other communities could even see a relation between x and B and how they affect each other but B would be unaware of it.

The existence of a given entity/system x (as perceived by community A) may afford real and present properties that are not dependent on the capacity of community B to identity them. Thus x exist somehow *independently of B*, but it does not exist *for B*. However, although unaware of that reality, depending on the position that B takes in relation to the configurations that sustain the existence of x (as perceived by A), it may alter or change this reality or experience changes as a function of it. This reality may, however, be more or less strongly sustained by A or other communities of observers and relations, that create affordances for other relations and couplings with other observers (Gibson 1986).

However, what the observers will be able to perceive is also dependent on their structural capacities (structural determination) and modes of coupling (Maturana and Varela 1992). No observer, except maybe a primordial one (emergent with the first act of distinction) can have access to complete knowledge and to all possible punctuations and distinctions. Knowledge is dependent on boundaries and constraints, associated with acts of distinction (Cilliers 2002; Kauffman 1987). We cannot know something except under the constraints set up: (i) by our own internal structure; (iii) by our contributions to the world, through our coupling with it; (iii) by what the organisation of the world, as sustained by given communities of observers and matrices of relations, affords as possibilities for our coupling with it, (iv) by the nature of the resulting coupling and its feedback and recursive influences (v) by the positive and negative constraints set up by the other relations in which we more directly participate, and a wider network of relations that sustain particular configurations of properties in the world.

We may now revisit the ideas introduced before to assist in clarifying a relative framing of the notion of complexity.

Depending on their structural determination (Maturana and Varela 1992) different observers, with different relative capacities will be capable of perceiving/constructing different realities and of engaging in different types of relation and coupling modes with other entities, hence giving rise to new realities and changing the landscape of possibilities for action differently, for themselves and others. This capacity relates to their complexity, but judgements about their complexity can only be made in relative terms. Under certain conditions, statements about complexity or about complex thinking can be made *as if* they were absolute. Nevertheless, one needs to be aware of the underlying assumptions and special conditions that allow for such treatment. Complexity is, necessarily, a relative concept and one needs to clarify which relative perspectives are used to compare differences of complexity (between and within observers, systems and couplings processes and outcomes). In certain conditions, some of these comparisons can be taken as proxies for others, and complexity can be treated *as if* it was an absolute property, but it remains important to clarify the assumptions on which this strategy rests.

Hence, when we aim to pursue knowledge or discover a reality "out there" we are talking, to some extent, of a reality that is set forth by our coupling with a world that is "out there" but one that assumes particular shapes only in those interactions. We are also pursuing knowledge constructed and held by other communities of observers or participants in a relational reality, either known or hypothetical, that, through their own shapings, create affordances for our own coupling. Reality exists as a relational matrix that sustains an array of different possibilities for its construction for its different participants. For some, reality can be more differentiated and integrated than for others that can only experience it from very limited perspectives. Our knowledge is always relative to our position in that network of relations. Complexity can be said to exist as a property of reality that is anchored in its relationality and the potential for differentiation, integration, recursiveness and emergence. But to talk about complexity as a property of the world implies assuming a relational worldview where the evaluation or realisation of complexity is relative to the position of the observer, of the systems of interest and of their coupling in a wider relational frame.

The complex world may exist independently of (particular) observers, but it does not exist in a state of absolute independence as it is organised and sustained by relations. Different configurations of relations sustain different expressions in the fabric of reality and different communities of observers, themselves sustained by patterns of relations, also participate in this fabric. Some communities of observers, due to their position in these networks of relations, may be capable of higher levels of integration, and thus may have a more salient role in sustaining or affording (Gibson 1986) a reality that is perceived by other (communities of) observers, from their particular positions, as 'given' to them. Some observers will be more complex than others to the extent that they are capable of couplings that create more differentiated and integrated/interconnected constructions of reality. They are able to adopt a wider diversity of perspectives and of coordinating and integrating them thereby generating a wider array of possibilities for action as well as modes of coupling that, as a consequence, produce more information/difference and eventually lead to changes in the networks of relations in which they are sustained. Some communities will be able to integrate and surpass the constructions of others. Aiming for "more" or "better" knowledge could be conceived as a goal that is defined in relation to a hypothetically more complex group observers or in relation to some previous state of a given observer of interest. Some observers (particularly e.g. in the human social domains) may act in ways that lead to deep transformations in their network of relations that significantly shape and alter their realities. But others, if not fully embedded and in a critical position in the network of relations that defines an object x, a given observer or entity may have an existence to which a particular x is relatively indifferent to.

Let us consider an Ideal(ised) Observer (IO) (an example would be as a primordial observer emerging from a mass of undifferentiated relations in the Universe). In principal, an IO is capable of participating in all possible relations with all other existing entities, either directly or indirectly, and of drawing an infinite number of boundaries and cuts in the relational matrix of the Universe, through all possible time frames, hence adopting an infinite number of perspectives, views or positions: in essence it has access to all knowledge. An IO should be capable of situating and identifying all other relational entities, systems, observers and communities of observers and positioning them in all possible configurations of relations. An IO should, therefore, be capable of comparing the modes of coupling and constructions of different communities of observers in terms of their relative complexity.

In Fig. 2.2 we illustrate an hypothetical stance of an Ideal (or idealised) Observer- eventually a primordial observer- in relation to other communities of observers.

For the communities of observers A, B and C the System of Interest x, appears with particular different contours and dimensions. The IO is capable of identifying all of these constructions but also of understanding that they are constrained by the relational contexts of the particular configurations of relations that delimit the boundary of, and thus the internal and external relations of, these communities.

The system X exists in a certain way for given community of entities and observers. When engaging with a certain system and trying to understand its complexity, other (more or less known or abstract) communities can be taken as reference points to consider what is currently known or could be known about the system. The relations that sustain that system, as identified by a given community of observers, create a landscape of possibilities and regularities in terms of boundary definitions and affordances for coupling that will allow that system to be recognised by others in similar terms. When an observer couples with the system in a way that is not only congruent with the relational organisation that sustains it, as perceived by a given community of observers, but also that is also coherent, in the sense it affords more differentiated and integrated descriptions, the situation may lead to increased possibilities for action. We propose that when the coupling is organised by a set of principles that we currently believe organise complex systems, then there is scope for new information to emerge that will increase the complexity of the coupling and

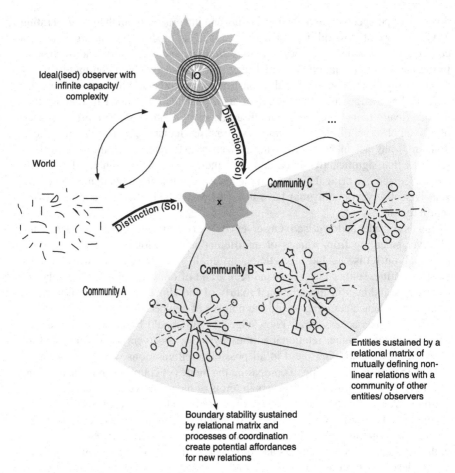

Ideal(ised) observer with
infinite capacity/
complexity

World

Distinction (SoI)

Distinction (SoI)

Community C

x

Community B

Community A

Entities sustained by a
relational matrix of
mutually defining non-
linear relations with a
community of other
entities/ observers

Boundary stability sustained
by relational matrix and
processes of coordination
create potential affordances
for new relations

Fig. 2.2 The Idealised Observer (IO) in relation to other communities of observers (A, B and C) and systems of interest (SoI)

thus for novel outcomes/information to emerge that themselves may further guide the coupling towards increased complexity and thus enhanced possibilities for action and change.

Hence understanding the nature of the coupling between an observer and a system of interest and the degree to which this coupling results in coherence and increases in complexity is fundamental to our understanding of how to gain understanding of complex systems that could support the management of change. Different observers will be constrained by their modes of coupling in their possibilities for action. In Fig. 2.2, the system-of-Interest appears with different contours and dimensions for the communities of observers A, B and C. The Ideal Observer should be capable of comprehending and articulating the constructions of all other possible observers, in the relational contexts where they emerge, as well as of assessing their positions in relation to a system-of-interest in terms of the relative complexity of their constructions, through all possible time slices (durations and frequencies).

Figure 2.3 presents an abstracted example of different types of comparisons that could be made between and within observers and systems and how one could issue statements about their complexity in relative terms. It exposes the conditions under which statements complexity could be made *as if* it was absolute and provides some examples.

The figure portrays the views of four different observers (or observer communities), Oa, Ob, Oc and an IO. Let us suppose that these observers are capable of identifying, for a given purpose, certain systems-of-interest. The columns represent the evolution of their views of that system, through time, but it can also represent their comparative views of different systems. The rows represents examples of different types of views or perspectives (P) as snapshots of configurations of relations that distinguish that system, according to a given observer, at a particular point in time. The geometric patterns represent different hypothetical dimensions: parameters through

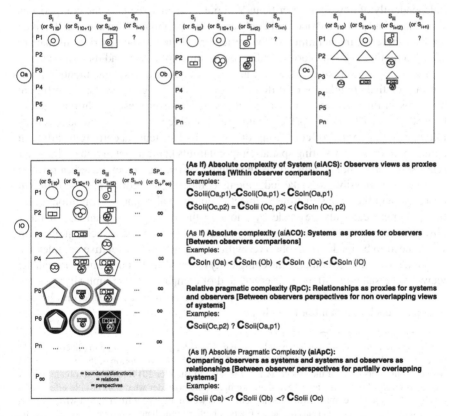

Fig. 2.3 Abstract representation and examples of comparisons of the complexity of systems and observers

which the observer is capable of describing or interacting with the system or different ways of drawing boundaries and distinctions; they represent dimensions of the differentiation of the thinking. The relative positions of these forms could represent abstract relations between these dimensions, and ways of relating and integrating them, for example, in some figures some dimensions encompass others, while in others they are placed at the same level.

The information corresponding to the Ideal Observer is but a sample of an infinity of perspectives that this observer could take; these naturally include perspectives that are more complex than those of other observers. Absolute statements about the complexity of the world, from the perspective of different observers can only be made in relation to the complexity of this IO. Therefore, even if they seem absolute, they are in fact relative to the hypothetical perspectives or views that this IO with infinite complexity.

In relation to this IO, it would be possible to make statements about different degrees of complexity of different systems and between different observers in relation to that system, as long as there is a minimum agreement on the boundaries of that system that would allow the different observers to recognise it as System X, independently of how they describe it. Without such an agreement the comparisons would make no sense and one could only assess the pragmatic relevance[3] and impact on their choice of the definition of the System in relation to the descriptions that are afforded by them and the corresponding possibilities of action and their effects.

With such an agreement it would also be possible to make statements regarding the relative complexity of different systems, or of a given system at different moments in time, within or between observers. However these conditions need to be made clear. The IO is capable of taking all possible times slices and perspectives into consideration and of comparing observers and systems according to different agreed boundaries. By setting spatial-temporal limits in the target region of the wider relational matrix, boundaries are defined and a system can be extracted in relation to which it is possible to make judgements comparing the complexity of different observers and their constructions (subject to consensual boundary conditions). But this judgment can only be made by adopting the perspective of an IO, or other observers more complex than those being analysed; we note that such a judgement would need to be made by an observer with sufficiently higher relative complexity.

It is also possible to compare the complexity of a system, through time, through the views of the same or different observers and of comparing the complexity between their views for an agreed system or set of systems of interest, considering them in relation to each other and/or to an IO.

---

[3]Cf. Pragmatism views on knowledge and truth: e.g. "consider what effects, that might conceivably have practical bearings, we conceive the object to have. Then, our conception of these effects is the whole of our conception of the object" (CP 5.388-410, In Buchler 2014, p. 31); "to obtain perfect clearness in our thoughts of an object, then, we need only consider what conceivable effects of a practical kind the object may involve- what sensations we are to expect from it, and what reaction we must prepare. On conceptions of these effects, whether immediate or remote is then for us the whole of our conception of the object, so far as that conception has positive significance at all. This is the principle of Peirce, the principle of pragmatism" (James 1955, p. 43).

The examples presented in the Fig. 2.3 support a simplified operational definition and examples of the conditions under which it is possible to make statements of *as if* absolute and relative complexity in four different types of scenarios:

1. *(As if) Absolute Complexity of System*: here observers views for (minimally agreed bounded) given systems are taken as proxies for statements about the complexity of the systems. There are scenarios in which Relative Complexity is treated as if Absolute Complexity of Systems when the multiple within observer perspectives and their relationships with the systems are taken as proxies for the complexity of the System. The complexity of the system is evaluated by considering the relation to the complexity of the observer's perspectives (in the context of their relationship to the systems). Observer perspectives and system collapse and are treated *As If* they were the same so that judgements can be made on the complexity of the latter *As If* it was absolute: *As If* the Systems existed outside of some observer perspective.

2. *(As if) Absolute Complexity of Observer:* here the systems and relationships as taken as proxies for Observers. In some situations, when comparing all possible observers' perspectives as a result of their coupling with the systems, there is at least some partial overlap between their perspectives and some may be capable of encompassing or integrating others, due to being more complex than them. In these cases (in conditions of minimal agreement on the definition of the system and its boundaries) we may talk about the Relative complexity of the observer *As If* it was Absolute, considering the observer capable of the most encompassing and integrative views as the most complex. Even though the system will pose different challenges and present different affordances, based on the history of couplings in its configuration of relations and with previous observers, their relationships with the present observers, and the resulting perspectives emerging from the coupling, are taken as proxies for the complexity of the Observers.

    This type of comparison considers that the most complex observer would be the one capable of the most complete knowledge and of the largest number of descriptions, distinctions and perspectives and, simultaneously, with the capacity to integrate them. All observers are also compared in relation to an IO.

3. *Relative Pragmatic Complexity of Observers*: here relationships are considered as proxies for observers and systems. In cases where there is only little or no overlap between the observer's perspectives and where none of them appears to be capable of integrating the others' perspectives (even to the point where they might appear to be different systems) then only judgments of Relative Pragmatic Complexity can take place. These judgements are also restricted to situations where there is at least minimal agreement concerning either the purposes or intentions of the coupling with the system (e.g. to promote a certain type of change) or the intentionalities of the system itself. In these situations it is possible to collapse systems and observers into their relationships, judging complexity in terms of the possibilities afforded by coupling process in relation to a given purpose/intention, assessing the fitness of the pragmatic outcomes in terms of their its effect and desirability. In Relative Pragmatic terms, the relationship that

affords more possibilities for action[4] and which leads to an increased number of choices that lead to more positive outcomes in relation to a given set of goals or values, could be said to be, the more complex.

4. *(As if) Absolute Pragmatic Complexity of the Observers:* here systems and relationships/couplings with observers are taken as proxies for the complexity of the observers. This is a combination of the conditions for (As if) Absolute Complexity of an Observer and Relative Pragmatic Complexity (cases 2 and 3 above) in situations where there is at least a minimal overlap between observer perspectives and where it is possible to compare the perspectives resulting from the coupling of the observer and the system, and take them as proxies for the Complexity of the Observer, considering the diversity and integration of their views in relation to one another. However, even though, in principle, an observer could be considered as more complex than another, by virtue of having a greater diversity of possibilities of actions available, ultimately it is the fitness of the outcomes of their choices that will determine their complexity. This fitness of outcomes is influenced by the coupling process and the purpose/goal that guides it as well by the internal organisation of the observer, their history and the relational context that creates particular pressures and constraints. Therefore, although a given observer could, in principle, be considered to present greater (As if) Absolute complexity or potential for complexity, they could be judged as less complex if, given a certain array of possibilities for action, their choices can be judged to result in poorer outcomes, by a given set of critical observers.[5] They may have failed to consider a wider set of relations and contextual conditions or even aspects related to their coupling that would, for example, lead to the choice of a simpler course of action that would be the best match to the complexity of the change scenario, considering a given goal or purpose.

References

W.R. Ashby, Requisite variety and its implications for the control of complex systems. Cybernetica **1**, 83–99 (1958)

G. Bateson, *Mind and Nature: A Necessary Unity* (Bantam Books, New York, 1979)

J. Buchler (ed.), *The Philosophical Writings of Peirce (originally published in 1955)* (Dover Publications Inc., New York, 2014)

D. Byrne, G. Callaghan, *Complexity Theory and the Social Sciences: The State of the Art* (Routledge, London, 2014)

---

[4]This position relates to Von Foerster's (2003c, p. 282) Constructivist Ethical Imperative stating "I shall act always so as to increase the total number of choices".

[5]By critical observers we mean all of those entities who are more or less closely engaged in the network of relations that creates or sustains a given system of interest (as identified by a particular set of observers) as well as those observers that may more or less directly or indirectly affect or be affected by change in such system of interest. To some extent the idea of a critical observer relates to the notion of stakeholder but that we propose is more strongly grounded in a relational worldview, bot being restricted to human observers.

J.L. Casti, On system complexity: identification, measurement, and management, in *Complexity, Language and Life: Mathematical Approaches*, ed. by J.L. Casti, A. Karlqvist (Springer, Berlin, 1986), pp. 146–173

J.L. Casti, A. Karlqvist, Introduction, in *Complexity, Language and Life: Mathematical Approaches*, ed. by J.L. Casti, A. Karlqvist (Springer, Berlin, 1986), pp. xi–xiii

L. Caves, A.T. Melo, (Gardening) Gardening: a relational framework for complex thinking about complex systems, in *Narrating Complexity*, ed. by R. Walsh, S. Stepney (Springer, London, 2018), pp. 149–196. https://doi.org/10.1007/978-3-319-64714-2_13

P. Cilliers, Why we cannot know complex things completely. Emergence 4(1/2), 77–84 (2002)

P. Checkland, J. Scholes , *Soft Systems Methodology in Action. Includes a 30-year Retrospective* (Wiley, Chichster, 1990)

P.A. Corning, The re-emergence of "emergence": a venerable concept in search of a theory. Complexity 7(6), 18–30 (2002)

B. Edmonds, Complexity and scientific modelling. Found. Sci. 5, 379–390 (2000)

J.J. Gibson, *The Ecological Approach to Visual Perception* (Psychology Press, New York, 1986). (Originally published in 1979)

J.A. Goguen, F. Varela, Systems and distinctions: duality and complementarity. Int. J. Gen Syst 5, 41–43 (1979)

H. Haken, *Synergetics* (Vieweg + Teubner Verlag, Wiesbaden, 1973)

F. Heylighen, *Complexity and Evolution. Fundamental Concepts of a New Scientific Worldview*. Lectures notes 2017-2018 (2018), Retrieved from http://pespmc1.vub.ac.be/books/Complexity-Evolution.pdf

F. Heylighen, P. Cilliers, C. Gershenson, Complexity and philosophy, in *Complexity, Science, and Society* (2006), http://cogprints.org/4847/

W. James, *Pragmatism and Four Essays from The Meaning of Truth*. (New American Library, New York, 1955). (Pragmatism originally published in 1907; The Meaning of Truth originally published in 1909)

L. Kauffman, Self-reference and recursive forms. J. Soc. Biol. Struct. 10, 53–72 (1987)

B.P. Keeney, *Aesthetics of Change* (The Guilford Press, New York, 1983)

S. Kelso, Coordination dynamics, in *Encyclopedia of Complexity and Systems Science*, ed. by R.A. Meyers (Springer, New York, 2009), pp. 1537–1564

S.J.A. Kelso, D.A. Engstrom, *The Complementary Nature* (MIT Press, Cambridge, MA, 2006)

J. Ladyman, J. Lambert, K. Wiesner, What is a complex system? Eur. J. Philos. Sci. 3(1), 33–36 (2013)

R.M. Lerner, *Concepts and Theories of Human Development*, 3rd edn. (Lawrence Erlbaum Associates, New York, 2002)

M. Lissack, The context of our query, in *Modes of Explanation. Affordances for Action and Prediction*, ed. by M. Lissack, A. Graber (Palgrave Macmillan, New York, 2014), pp. 25–55

J. Macy, *Mutual Causality in Buddhism and General Systems Theory: The Dharma of Natural System* (State University of New York Press, New York, 1991)

S.M. Manson, Simplifying complexity: a review of complexity theory. Geoforum; J. Phys. Hum. Reg. Geosci. 32(3), 405–414 (2001)

H. Maturana, Biology of language: the epistemology of reality, in *Psychology and Biology of Language and Thought. Essays in Honor of Eric Lenneberg,* ed. by in G. Miller, E. Lenneberg (Academic Press, Cambridge, MA, 1978), pp. 27–63

H. Maturana, F. Varela, *The tree of Knowledge. The Biological Roots of Human Understanding* (Shambhala, Boston, MA, 1992)

L. McIntyre, Complexity: a philosopher's reflections. Complexity 3(6), 26–32 (1998)

D.W. McShea, R.N. Brandon, *Biology's First Law: The Tendency for Diversity and Complexity to Increase in Evolutionary Systems* (University of Chicago Press, 2010)

C.R. Mesle, *Process-Relational Philosophy. An Introduction to Alfred North Whitehead* (Templeton Press, West Conshohocken, PA, 2008)

E. Morin, From the concept of system to the paradigm of complexity. J. Soc. Evol. Syst. **15**(4), 371–385 (1992)

S. Oyama, P. Griffiths, R. Gray, *Cycles of Contingency. Developmental Systems and Evolution* (The MIT Press, Cambridge, MA, 2001)

M. Reynolds, Evaluation based on critical systems heuristics, in *Using Systems Concepts in Evaluation: An Expert Anthology*, ed. by B. Williams, I. Imam (EdgePress, 2007), pp. 102–122

M. Reynolds, Bells that still can ring: systems thinking in practice, in *Moving Forward with Complexity: Proceedings of the 1st International Workshop on Complex Systems Thinking and Real World Applications*, ed. by A. Tait, K. Richardson (Emergent Publications, Litchfield Part, AZ, 2011), pp. 327–349

M. Reynolds, S. Holwell (eds.), *Systems Approaches to Managing Change: A Practical Guide* (Springer, London, 2010)

K. Richardson, The hegemony of the physical sciences: an exploration in complexity thinking. Futures **37**(7), 615–653 (2005)

R. Rosen, Complexity as a system property. Int. J. Gen. Syst. **3**(4), 227–232 (1977). https://doi.org/10.1080/03081077708934768

R. Rosen, *Life itself. A Comprehensive Inquiry into the Nature, Origin and Fabrication of Life* (Columbia University Press, New York, 1991)

R. Rosen, *Essays on Life Itself* (Columbia University Press, New York, 2000)

A. Stetsenko, *The Transformative Mind: Expanding Vygotsky's Approach to Development and Education* (Cambridge University Press, New York, 2017)

G. Tononi, O. Sporns, G.M. Edelman, A measure for brain complexity: relating functional segregation and integration in the nervous system. Proc. Natl. Acad. Sci. U.S.A. **91**(11), 5033–5037 (1994)

W. Ulrich, Reflective practice in the civil society: the contribution of critically systemic thinking. Ref. Pract. **1**(2), 247–268 (2010)

F. Varela, Not one, not two. *The Coevolution Quarterly, Fall* (1976), pp. 62–67

H. Von Foerster, On constructing a reality (Address originally published in 1973), in *Understanding Understanding: Essays on Cybernetics and Cognition*, ed. by H. Von Foerster (2003) (Springer, New York, 2003a), pp. 211–227

H. Von Foerster, Cybernetics of cybernetics (originally published in 1979), in *Understanding Understanding: Essays on Cybernetics and Cognition*, ed. by H. Von Foerster (2003) (Springer, New York, 2003b), pp. 283–286

H. Von Foerster (2003c). Order/Disorder. Discovery or invention? (originally published in 1984), in *Understanding Understanding: Essays on Cybernetics and Cognition*, ed. by H. Von Foerster (2003) (Springer, New York, 2003c), pp. 273–282

A.N. Whitehead, *Process and Reality (corrected edition)* (The Free Press, New York, 1978)

M. Williams, W. Dyer, Complex realism in social research. Methodol. Innov. **10**(2), 1–8 (2017). https://doi.org/10.1177/2059799116683564

Chapter 3
Complex Thinking as the Complexity of the Coupling Observer—World: Processes and Outcomes

Abstract In this chapter, building on the work of Edgar Morin and on a relational worldview, we conceptualise Complex Thinking as a mode of coupling with the world, understood as both process and outcome. The creation of knowledge is conceived as resulting from the differences generated in the coupling relationship between an Observer and the world. We introduce a proposal in which Complex Thinking is conceptualised and operationalised through a set of dimensions and corresponding properties inspired by the characteristics of the natural, biological and social worlds. We discuss how the enactment of these properties allows an Observer to manage their contributions to the coupling relationship, increasing its complexity and coherence in a way that may lead to the production of new and meaningful information to further guide their actions towards (more) positive outcomes. We distinguish between first and second-order Complex Thinking and discuss their relationship with abductive forms of thinking and their roles in guiding action in situations of uncertainty and ambiguity. We discuss our definition of Complex Thinking in relation to the notions of Complexity Thinking and Complex Systems Thinking, both differentiating and connecting them.

Keywords Modes of coupling · Complex thinking · Abduction · Abductive thinking · Complexity thinking · Complex systems thinking

3.1 Complex Thinking as a Mode of Coupling

Building on the work of Morin (2005) and situated within a relational constructive worldview, we conceive of Complex Thinking as a mode of coupling defined both as (i) a coupling process (mode of coupling) and (ii) an outcome of that coupling process. This conception is influenced by a relational and enactive view of cognition, where cognitive activity is conceived as "a history of the structural coupling that brings forth a world" (Varela et al. 1991, p. 206) where observer and world co-exist in a relation of mutual determination.

Our modes of thinking or coupling with the world generate differences/information that shape the way that we further interact with the world and (co-)evolve with it. As a mode of coupling, complex thinking can produce information through the creation of differences in: (i) one's own states in relation to the world/system; (ii) the emergent view of the world, and the system-of-interest, in relation to our own previous views or those of other observers; and (iii) the organisation of the relationship between the observer and the world and the experience of its effects. The creation of this information may lead to changes in the thinking and in turn lead to more differentiated and integrated perspectives, offering more possibilities for effective action through more congruent couplings.

A complex organisation of relations is an essential corollary of complex thinking (Morin 1992). To face the challenges of the world and of its complexity it is not enough to think *about* complexity: it is critical to think *in* complexity. Hence, in this proposal, Complex Thinking is as much the *outcome* of a type of relationship with the world, as it is a *relational practice or process* that assumes the form of a "lived" complexity (Chia 2011; Rogers et al. 2013).

Complex Thinking thus encompasses both:

1. A *mode of coupling: complex thinking* as a *process that is sustained by a set of practices* that simultaneously: (i) attend to (describing, explaining, predicting) and adjust to the complexity of (a selected part of) the world (the system of interest) and the properties that sustain its complexity (as recognised by given communities of observers at a given point in time); and (ii) enact such properties as contributions to the coupling relationship;
2. An *outcome of coupling:* complex thinking generates (i) a multiplicity of descriptions, explanations and predictions as well as framework for their integration; (ii) meaningful emergent novel information, translated as differences that makes a difference (Bateson 1979) in the observer, the target system and/or their coupling relationship towards increased coherence and complexity; (iii) a variety of possibilities of action for promoting, supporting or managing change in both the observer, the world, and their subsequent coupling relation, guiding choices that build (iv) constructive interactions and positive co-evolving relationships capable of sustaining positive outcomes for the observer, the target system and their environments, as agreed by a set of critical observers (entities either involved and/or more or less directly affected by the outcomes).

In the proposed conception of complex thinking *both* process and outcome need to be considered in evaluating the complexity of the thinking.

While more complex forms of thinking will generate a wider array of alternatives for action, they should also lead to pragmatically effective choices and positive changes (as judged by a set of critical observers). Because the complex integrates the simple, the most effective choice could be associated with a very simple course of action, but one that is more likely to be ecosystemically fit (Caves and Melo 2018) and more congruent with the dynamics of the given ecology of actions in which it is embedded (Morin 2014). The complexity of the thinking, as an outcome of the coupling, could also be assessed in terms of the accuracy of the projections and visions it generates and the consequences of the actions it informs. Hence, there is an important pragmatic dimension to the evaluation of complex thinking as an outcome.

We have introduced Complex Thinking as a relational concept where the degree of 'complexity' is understood in relative terms. However, it is also, necessarily, a dynamic, evolving and context-dependent concept to the extent that our knowledge of the world and our conceptions of complexity are likely to continue to evolve towards increasing differentiation and integration leading, in a recursive fashion, to the emergence of new ideas, patterns, descriptions, explanations and worldviews that we currently might not be able to anticipate.

3.2 Dimensions and Properties of Complex Thinking

Our proposal for complex thinking follows the contributions of Morin in defending a type of thinking that is congruent with the processes that are deemed to sustain the complexity of the world. We seek inspiration in the properties of the complex world to (re)think how we may organise and practice our thinking in ways that may better capture as well as enact that complexity. As Morin stated "complexity is also a mode of knowledge when we integrate certain principles: the principle of retroactivity, of connectivity, in a dialogical principle. It is a way of thinking" (Morin 2014, p. 19).

A key assumption of our proposal is that when the thinking is organised according to principles and properties similar to those that organise the complex world then it is more likely to exhibit emergent properties and capacities that expand our possibilities for actions that are more likely to be ecosystemically fit and sustainable. Although we agree with Morin's statement that all systems, by virtue of their relational organisation, should be considered complex, we also accept the use of the expression "complex systems" to refer to those systems of interest in relation to which some expressions of (relative) complexity are particularly evident, not only in terms of their high(er) differentiation and integration, but also in terms of the recursive processes that are associated with their eco-self-organising dynamics (Morin 1992) and their capacity to generate novelty and surprise their observers.

The novelty generated by complex systems in the coupling with their environments, adds to their own differentiation and integration while making contributions to their coupling dynamics and degree of coherence (Maturana and Varela 1992). To deal with this novelty the observer needs to respond in coherent ways and co-evolve with the target systems by coordinating with it. Moreover, we propose that complex thinking relates to both the process and outcome of coupling. The complexity of the thinking, as an outcome, can be enhanced through the enactment of a thinking process that embodies critical properties of systems that, as expressions of their complexity, are capable of changing and co-evolving with their environments, generating the necessary novelty to adapt with changes that build robustness and resilience in the face of internal and external challenges.

The capacity of the observer to make positive contributions to the coupling relationship with the target system-of-interest will relate to their own (relative) complexity and the extent to which their thinking is differentiated, integrated and capable of learning and evolving through cycles of recursion, leading to the emergence of new elements (ideas, hypotheses, patterns, processes) that adds capabilities that serve to enhance their capacity for increasing the congruence of their coupling relationship, in a recursive fashion.

The more complex the systems-of-interest, the more likely they will generate reactions of surprise in their observers (McDaniel and Driebe 2005) who may experience their (relative) complexity as difficulties in understanding, prediction, control or management (Edmonds 2018). Thus, while many of these systems challenge our complexity as human observers, resulting in frustrations and difficulties in our (attempts at the) management of our relationship with them, they also elicit feelings of awe, wonder and surprise.

The creative capacity to generate novelty, evident in numerous natural and social systems, lies at the core of the complexity of the world. Surprise and novelty call for particular types of thinking and exploratory approaches that are themselves capable of generating something new and of evolving and adapting in the relation with the world and the target systems of interest.

Abductive reasoning is, to some extent, a form of creative thinking that organised in a logic of discovery leads to the formulation of new explanatory hypotheses.[1] Charles Sanders Peirce (CP, 6.522–8, In Buchler 2014) presented the following as generic forms of abductive inferences:

> A surprising fact C is observed,
>
> But if A were true, C would be a matter of course
>
> Hence, there is reason to suspect that A is true" (CP, 5.189).

> A well-recognized kind of object M1, has for its ordinary predicates P1, P2, P3 etc., indistinctly recognized.
>
> The suggesting object, S, has these same predicates P1, P2, P3
>
> Hence, S of the kind M
>
> (CP, 5.542, 544–5 and James, 1903)

> M has, for example, the numerous marks P', P'', P''', etc.,
>
> S has the proportion r of the marks P', P'', P''', etc.
>
> Hence, probably, and approximately, S has an r-likeness to M
>
> (CP, 2.694–7).

Abductive (Fann 1970; Melo 2018; Nubiola 2005) and imaginative (Whitehead 1978) leaps are, in themselves, likely expressions of the complexity of our human minds and of our relationship with the world.[2]

---

[1]"A man must be downright crazy to deny that science has made many true discoveries. But every single item of scientific theory which stands established today has been due to Abduction" (CP, 7.172, 903; Cit in Nubiola 2005, p. 126).

[2]"Peirce appeals to the affinity between mind and universe (…) 'Our minds having been formed under the influence of phenomena governed by the laws of mechanisms, certain conceptions entering into those laws become implanted in our minds, so that we readily guess what the laws are. Without such a natural prompting, having to search blindfold for a law which would suit the phenomena, our chance of finding it would be as one to infinity (CP 6.10, 1891)'" (In Nubiola 2005, p. 127).

Creativity and abducting are closely related concepts. Creativity, defined as "the ability to come up with ideas and artefacts that are new, surprising and valuable" (Boden 2004), may lie "essentially in the way in which the subject relates the elements available in different realms of his and her experience" (Nubiola 2005, p. 126).

Abductive thinking (Fann 1970), operating within the logic of surprise (Nubiola 2005), can be integrated in a set of abducting practices (Melo 2018) that may support the development of modes of thinking congruent with critical of complexity and its inherent creativity.

The concept of abducting was previously proposed to correspond to a meta-methodological practice (Melo 2018) that integrates the logic of abductive reasoning with an exploratory stance (of curiosity and openness) and a set of relational practices aimed at eliciting the emergence of new information, as well as the development of a creative relationship with the information available, in order to foster its expansion, reorganisation and transformation. The concept was advanced in the context of a discussion of interdisciplinary methods due to the challenges posed by complexity to our modes our knowing (Klein 2004; Melo and Caves 2020).

In the domain of thinking, as a process of coupling with the world, emergence is associated with abduction and with a type of novelty that opens new possibilities for action and transformation. As Peirce stated "The act of abductive suggestion comes to us like a flash. (…) It's true that the different elements of the hypothesis were in our minds before; but it is the idea of putting together what we had never before dreamed of putting together which flash the new suggestion, before our contemplation" (CP, 5.157, 181, 184–5, 1891).

We hypothesise that, in certain conditions, the coupling relationship of an observer to the world will lead to the emergence of sufficient and meaningful information to guide effective positive actions and the management of change, in the relation with a target system, that is effective in the absence of full information.

We propose that a strong, positive and sufficiently complex coupling, organised according to particular properties, is capable of guiding the coordination of the observer's actions with the target system towards increased coherence; through its recursiveness, this process supports learning and co-evolution.

Thus, complex thinking embraces abducting strategies or heuristics that support the observer in responding to the novelty and surprise in the world through a type of coupling that attends to some critical features of complex systems that shapes the creation and (re)organisation of the resulting information in a way that offers possibilities for action and emergent understanding more likely to result in positive outcomes.

Morin (1992, 2005, 2014) has proposed that complex thinking needs to integrate three core principles of complexity: (i) the *dialogical principle*, through which duality is maintained in the context of unity" (Morin 2005, p. 99); terms and phenomena that are complementary or antagonistic can be related and integrated, while differentiated; (ii) the *principle of organisational recursion or recursivity*, which allows for a process to be simultaneously a producer and a product of itself as part of the self-organisation of living systems and "significant for understanding complexity at the human level" (Morin 2014, p. 17); and, finally, (iii) the *hologrammatic principle* through which the thinking of the whole is represented and contained in the parts much as the parts in the whole, where the parts and the whole and considered as being simultaneously both more and less than each other (Morin 2014).

In contrast to the principles of disjunction and reduction stemming out of the dominant paradigm of simplification, complex thinking should be organised around "principles of distinction, conjunction and implication" (Morin 2005, p. 104). Complexity implies both differentiation and integration. Hence, the thinking must also be able to distinguish and relate different dimensions of reality.

Following Morin's insights, we propose that complex thinking must, on the one hand, attend to a particular, yet diverse, set of principles and properties of the complex world (as known by the observers in a given time). On the other, in order to create conditions for emergence, it must enact such complexity, by integrating those principles and properties in its own organisation (in order to expand the knowledge available to the observer and/or guide them in a positive and viable way in the absence of complete knowledge). To some extent complex thinking is a strategy to manage (relative) ignorance and uncertainty and to guide action in such conditions.

Standing upon the shoulders of Morin, and integrating contributions aligned with the practice of Systems Thinking and the study of natural, living and social systems under the umbrella of Complexity as a broad scientific domain of investigation, we present a pragmatic proposal for an operational definition of complex thinking in terms of a set of organising dimensions and properties. A list of dimensions and key properties is presented in Table 3.1.

The properties will be further elaborated in the following chapters.

The definition of each property will be elaborated in the next chapter. Some of these properties apply more to complex thinking as a process and others to complex thinking as an outcome. Nevertheless, most of them, through their own recursive organisation (being both product and producers of themselves), and their interdependence with other properties, will simultaneously characterise complex thinking as both a process and as outcomes, that will further constrain and shape the processes (in a recursive manner). We conceive of these properties as organised in a type of network structure, influencing each other in non-linear ways. The nature of the structure and dynamics of this network calls for further investigation. It is likely that some properties will be more closely coupled and that some cannot built without others. It is also likely that some will act as facilitators or activators for others. Knowing the nature of the relations between these properties will provide important insights for the design of interventions aimed at promoting the complexity of the thinking.

Additionally we expect that the interaction between properties will lead to abductive leaps that may play a central role in the identification of critical processes that attend in the management of the relationship with a target system. This dimension will be further explored in the next section in relation to the notion of emergent complex thinking.

It is important to note that the operational definition of the concept of complex thinking is likely to evolve as the complexity of the world evolves towards increasing differentiation and integration, and thus our complexity, as human observers, in relation to it. The concept is therefore not only relative but dynamic and evolving.

Table 3.1 Dimensions and properties of complex thinking

Dimensions of the complexity of the thinking	Properties
A. Structural complexity	1. Structural variety and dimensionality 2. Relationality 3. Recursiveness
B. Dynamic/process complexity	4. Timescales 5. Dynamic processes 6. Relativity, ambiguity and uncertainty
C. Causal and explanatory complexity	7. Modes and finalities 8. Historicity 9. Complex circularity 10. Emergence
D. Dialogic complexity	11. Dualities and complementary pairs 12. Trinities and levels
E. Observer's complexity	13. Multipositioning 14. Reflexivity 15. Intentionalities
F. Developmental and adaptive complexity	16. Developmental-adaptive value 17. Developmental evolvability
G. Pragmatic complexity	18. Pragmatic value 19. Pragmatic sustainability
H. Ethical and Aesthetical complexity	20. Ethical value 21. Aesthetical value
I. Narrative complexity	22. Differentiation and integration 23. Identities 24. Flexibility/openness

Therefore it is likely that the proposed list of properties will undergo several transformations. For the present, the description of a set of dimensions and properties of complex thinking affords an operationalisation of the concept of complex thinking that may provide concrete guidance to its practice.

Many of these properties, and the practices that enact them are well developed, for example, in the systems thinking tradition (Reynolds and Holwell 2010). Some have been particularly highlighted by Morin and others relate to known properties of complex systems elaborated by complexity sciences. Nevertheless, congruent with a complex stance, we propose that the complexity of the thinking lies in none of these properties alone. It is, on the contrary, dependent on how much they contribute to the differentiation and integration of our constructions, in relation to a target system, to the strength of that coupling and to the extent that the combination of different properties leads to emergent understanding and possibilities for action towards positive change.

3.3 First and Second-Order or Emergent Complex Thinking

In many situations the information available about a target system of interest affords only very crude mappings of their features/behaviours (too partial or incomplete to support effective predictions) and the timeframes required for more detailed ones is incompatible with the rhythms of the pragmatic demands of the world; this generates contexts of high uncertainty and, oftentimes, of high risk (Funtowicz and Ravetz 1994). Complex thinking is especially required in these situations, in the form of second-order or emergent thinking. Building mappings of the world is, oftentimes, like building a puzzle where all the pieces are scattered or even missing.

In the process of assembling a puzzle different strategies can be adopted. For example, one can use a piece-by-piece, linear strategy, operationalised in search for the "right next piece" or alternatively one can adopt a trial and error strategy. Both of these approaches may eventually be successful, although not necessarily within useful time frames (or without some exasperation); the amount of resources necessary to find the missing pieces may be excessive or simply not available.

Sometimes, however, a single piece or a small set of pieces placed can be all that is needed when they provide a hint into the organisation of the whole, signalling the nature of the surrounding missing pieces and their configurational arrangements; this allows one to make sense of an otherwise dispersed, unintegrated or unclassified array of information. One acts on the assumption that "if this piece was correctly placed here, then those other pieces would make sense and aggregate in such and such a way". A single piece can then be the key towards assembling the whole by producing indirect information about it, in the form of hypotheses that guide further action (e.g. deciding for which type of pieces to look for next); consequently additional information can be produced that will support further mappings and a deeper understanding of the puzzle-system. Oftentimes, it is the outcome of (trial) actions informed by these hypotheses that produces (more) complete or critical information. However, those actions were guided by only partial information that, nevertheless was managed in a way that allowed for the emergence of novel ideas or hypotheses. This is the kind of indirect mapping, or second-order and emergent thinking that has propelled human knowledge, including the major leaps in Science.[3]

More complex situations call for a type of complexity in the coupling that leads to this form of *second-order or emergent complex thinking* realised in the form of abductive and imaginative leaps (Fann 1970; Whitehead 1978). These leaps, which

---

[3]As opposed to the incremental progress following such leaps, that are often associated with more deductive reasoning.

carry with them emergent, novel and pragmatically relevant information, oftentimes take the form of hypotheses and some type of abductive reasoning (Fann, 1970). This information may be used, heuristically, to probe the system in the context of a tight coupling with it and to 'fill in the gaps'. Using Whitehead's (1978) metaphor of imaginative leaps,[4] this kind of thinking resembles the movement of an airplane that takes off from solid ground, where evidence is found and information is created about the system and that, propelled by the principles of complex thinking, finds conditions to take a speculative jump. The airplane then needs to land again to make a new exploration of the ground terrain using the newly formed lenses.

The observer without having to directly match or fully map the complexity of a target system may, by managing the coupling processes, create sufficient complexity for the emergence of novel information or insights that may further not only reveal new dimensions but also integrate the information available before, without being directly reducible to it.[5] These novel insights, in the form of hypotheses, may be used to guide further actions in relation to the system that will act as probes bringing forth dimensions and processes that were not yet fully known or understood.

The emergent hypotheses can take the form of abductive inferences of the form "The Target-System X behaves in A ways and interacts with the Observer O that behaves in B ways. Their coupling leads to X changes, O changes and Z other changes. If C (hypotheses) were true, A and all these changes would be a matter of fact". Therefore, the emergent hypotheses that are created through a process of deep immersion and relation with the system through the modes available for complex thinking, will make sense of the system and of their changes, guiding further actions. In this process, new information can be sought that may confirm or reject the hypotheses C, but also bring forth new dimensions. The emergent hypotheses may also help to organise the available information and lead to a new cycle of direct exploration and rehearsals of actions that may clarify critical processes. First and second order complex thinking are tightly connected as the latter occurs during the process of the former under critical conditions. As cycles of direct (or first order) and indirect (or second order or emergent) thinking take place, new dimensions of the system may be brought forth.

Figure 3.1 illustrates the practice of first and second order complex thinking.

The top left side of the picture represents the enactment of different properties of complex thinking that construct distinct lenses that allow for particular types of distinctions and the construction of particular views on the system of interest. Each of these lenses per se may build a more or less differentiated and integrated view.

---

[4]"The true method of discovery is like the flight of an aeroplane. It starts from the ground of particular observation; it makes a flight in the thin air of imaginative generalization; and it again lands for renewed observation rendered acute by rational interpretation" (Whitehead 1978, p. 5).

[5]"If we give up trying to control, we gain access to enormous amounts of variety which vastly exceed any variety we might ever aspire to. This gives us the richest pastures in which to graze: an endless source of variety offering us insights and understandings we would never have otherwise had. This is a source of (individual) novelty and renewal, and hence is a potential source of creativity. In this view, one way of increasing creativity is by stopping trying to control, to manage, and enjoying the insights that this rich store of variety can offer us" (Glanville 2004, pp. 7–8).

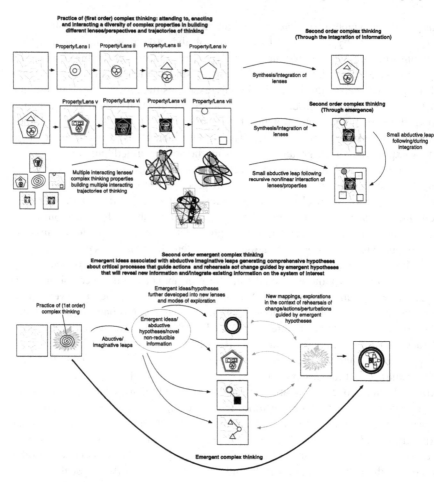

Fig. 3.1 Abstracted representation of the process of first and second order complex thinking

As we overlap lenses it may be possible to achieve some degree of synthesis and integration. The top part of the figure also illustrates how both this type of synthesis and integration and the intentional combination and interaction of properties (along with the views/perspectives that result from their enactment) in multiple trajectories of thinking, can lead to abductive leaps and to the emergence of new information and hypotheses that cannot directly be traced to the information available before.

The bottom part of the figure further illustrates the process of second-order or emergent complex thinking, highlighting how the emergent ideas, that might have resulted from the interaction of different properties of thinking and resulting perspectives, in the form of abductive hypotheses, are used as heuristic devices for the generation of new lenses and strategies to further organise the coupling process and explore the phenomena of interest in a way that may generate more information.

In complex thinking the target system contributes to the complexity of the observer through their coupling; the observer/intervenor manages this process in order to improve its informational potential, whilst allowing themselves to be informed. In this process both the observer/intervenor and the target system are subject to change and the landscape of possibilities of action available to them will be more or less transformed, to the extent that their own structural determination and initial conditions allow. Hence, the information produced by the coupling will, necessarily, be available, in different ways, for different observers, based on their own individual complexity. The information available through the coupling will be available in different ways for different types of participants depending on their potential and capacity to making different types of distinctions. While the success of first order complex thinking may be dependent on the nature, but also on the amount of information produced on the system, second order thinking may rely more strongly on how the dynamics of the coupling process, and on how the available information is managed and allowed to interact. When the thinking occurs as strong coupling (Caves and Melo 2018) and has, itself, a complex organisation, it holds the potential for the creation of novelty and the emergence of information through a process that cannot be analytically analysed or broken down into pieces, because it results from a synthetic emergent movement. In a sense, the complex thinking/coupling allows the system observer—system-of-interest to inform the observer through abductive and imaginative leaps. The subsequent cycles of action—information, construction—reconstruction of the thinking, will also likely allow the observer to more know about themselves in the context of their interaction and to increase their own complexity.

We suggest that there might be a minimum set of conditions under which first order complex thinking leads to second-order or emergent complex thinking so as to produce meaningful and pragmatic relevant information to support the management of the change processes in relation to a target system. We expect some conditions to be of special importance, namely, that:

(i) there is a minimum of (system-dependent) information available to build a preliminary relational view of a system-of-interest, with a focus on the coupling processes, namely information regarding (as if) internal coupling/complexity, coupling with the environment (as defined by a set of boundary distinctions) and coupling with the intervenor/interventions and to their own complexity as contributions to the coupling (Caves and Melo 2018);

(ii) there is a minimum of variety of types of information created by coupling in different domains (e.g. in social systems coupling both at cognitive, emotional, physical level) and with different sources;

(iii) there is a strong coupling with the system of interest and/or its environment so that the observer can quickly update their knowledge regarding the status of the system, monitoring minimal changes;

(iv) the coupling process integrates and enacts a variety of 'complex properties' that may vary in importance for different domains and different types of systems-of-interest; and

(v)   the information produced by different properties of the coupling processes is allowed to interact dynamically and recursively.

These conditions might assume different contours for different types of systems. The last two conditions pose the challenge of assembling a variety of heuristics, strategies and tools to promote a particular set of coupling properties. A variety of strategies, tools and heuristics are likely already used in different domains, that attend to and enact particular properties, namely in the domains of Systems Thinking in Practice (e.g. Checkland and Scholes 1990; Reynolds and Holwall 2010) and Complexity Science tools (e.g. Bar-Yam 2004). Their mapping, in relation to a framework of dimensions and properties for complex thinking, could assist in developing strategies to promote and assess the practice of complex thinking, for a given purpose. Additionally, a pragmatic theoretical framework for complex thinking can take the role of a meta-framework where a diversity of contributions could be coordinated and integrated.

To embark on emergent complex thinking the observer needs to abandon illusions or aspirations of control and be willing to accept uncertainty and partiality of information. They may need to act more as gardeners (Caves and Melo 2018) or farmers (Edmonds 2018) and engage in more flexible modes of coupling and timescales that allow for their co-evolution in relation to the reality being investigated/constructed. Their views and their own complexity will likely evolve more by steered the coupling process and their own contributions to it than trying to control a target system.

This form of complex thinking is, we believe, a true embodiment of complexity. It is a process and a practice as much as an outcome that is sustained by a set of organising principles. A relational worldview (e.g. Caves and Melo 2018) may, nevertheless, inform more than propositions or factual statements and not by propositions or statements about systems, albeit it might include them, for example, by setting an initial map or worldview on the basis of which the system is explored suggest a general relational map and a general framework to support this kind of exploration).

The thinking corresponds to the embodied cognitive activity of a knower that is subject to contextual constraints. The context of knowing is necessarily bounded and limited by the relational matrix that sustains the activities of the thinker. All knowledge is subject to boundaries (Cillers 2002) and constraints that are as much restricting as enabling to the extent that they produce differences that make a difference (Bateson 1979).

Albeit we focus on complex thinking as a form of purposeful activity, it might unfold in a less purposeful manner as part of the natural behaviour of coupled systems. Some observers may use less formal or intentional strategies yet, nevertheless, show great capabilities in the management of natural and social systems. An example would be the difference between the wise, experienced but (formally) uneducated master gardener and the highly educated but less experienced technical engineer (Caves and Melo 2018).

In any case we believe that complex thinking can be nurtured and improved through practice. We also contend that a clarification of the concept may allow for its operational development in the context of a pragmatic framework, where different dimensions and properties may be targeted and related towards increased complexity, as conceptualised from a relational point of view.

3.4 The Processes and Contents of the Thinking: Complex Thinking, Complexity Thinking and Complex Systems Thinking

A sufficiently complex observer could eventually (in the limit of an Idealised Observer) achieve a direct and full mapping of a target system. The mapping of complex systems can be assisted by the knowledge available about the behaviours of very complex systems, as well as their organising dimensions and properties. These are some of the critical contributions of Complexity Science as it progressively expands our knowledge of critical properties of the systems that appear to us as most complex.

This knowledge base is likely to evolve and be transformed. At any given time, it can be used to support some mapping process by informing the *contents of the thinking or when used as a heuristic lens to guide the exploration of a system of interest.*

Additionally this knowledge based can be used to inform the enactment of complexity in the observer's contribution to the coupling relationship with target systems, namely the properties that the *coupling process* could attempt to integrate or embody. To a large extent these two complementary ways of using the available knowledge about complexity and complex systems relate to the distinction we draw between Complex Thinking (associated with Morin's *general* complexity) and what oftentimes appears in the literature as *complexity thinking* or *complex systems thinking* (more often associated with Morin's *restricted* complexity). These expressions are sometimes used interchangeably and oftentimes are applied to different concepts and dimensions of thinking (in) complexity (Melo et al. 2019). We choose to reserve the expressions *complexity thinking* and *complex systems thinking* to refer mostly to the contents of the thinking that are informed by what has been called 'complexity theory' or 'complexity science'. Complex thinking is assumed as a broader concept, more focused on the processes and outcomes of the thinking, that may or may not explicitly integrate particular contents such as that considered in complexity theory and complexity science; it will, nevertheless, attend to many of the properties they call attention to. *Complexity thinking* may be used as an expression to refer to a conceptual armoury (Melo et al. 2019) as well as set of assumptions and theoretical statements that configure a way of making distinctions and punctuations in reality that highlight a set of processes and properties currently associated with complexity and complex systems (Cilliers 1998; Byrne and Callaghan 2014; Érdi 2007; Kelso 1995; Manson 2001). *Complexity thinking* uses complexity theory as a way to tune into critical features of these systems and to grasp their modes of operation (Richardson 2005, 2008). Albeit *complexity thinking* may invite a focus on the processes, we assume it mostly corresponds to the contents of the thinking and to a specific vocabulary. However, in some circumstances these contents, can be used as lenses that shape our relationship with a given reality and highlight particular features, to the relative neglect of others (Melo et al. 2019). They can, therefore, also be used as heuristics to organise information by attending to particular types of relations and distinctions, much as the tradition of systems thinking has proposed. Because the contents may pertain to processes, this type of thinking makes a contribution to what we call complex thinking. When understood at the level of the contents, complexity thinking may be taught and transmitted, revised and updated. However, the process that is enacted in the coupling with a target system is not necessarily congruent with, nor does it enact, the properties that it looks for in the world. We propose that this is the major point of distinction between *complex thinking* and *complexity thinking*, the former a larger and more general concept that can encompass the latter. *Complexity thinking* may or may not lead to a complex process of thinking, when it is more than a lens and attempts to be congruent with the complex world in way that leads to complex outcomes, and *complex thinking* may or may not use the lenses and contents of *complexity thinking* (Melo et al. 2019).

As a process, complex thinking attends to particular ways of looking at a system-of-interest and applies lenses that highlight the complex properties of the world. But its most defining feature is that it implies a practice of enactment of particular properties of complex systems. As a process relying on a set of practices, it can be improved, nurtured, enriched, trained and scaffolded but not taught in the stricter sense of the word. While complex thinking relates to a set of practices that sustain a process of coupling, it does not necessarily determine the contents of the thinking, albeit it might invite an observer to look for particular types of information. Hence, it may or may not integrate the contents of what we call complexity thinking. In fact, complex thinking could, in principle, be practiced intuitively, without any theoretical background, operating as, or akin to, other ways of tacit knowing, notions of wisdom (Baltes and Smith 2008). We propose that *complex thinking* is a broader concept than *complexity thinking* and albeit it might be enriched by it, it is not limited to it. *Complex thinking* can also be seen as a fundamental way to approach the expansion and development of *complexity thinking* and *complexity theory*, as it creates conditions for the emergence of new information about particular categories systems-of-interest considered as complex. *Complexity thinking* may allow us some knowledge 'about' a target system but it does necessarily build the type of understanding that is associated with a sense of coherence in our modes of explanation and practices (Lissack and Graber 2014). *Complex thinking*, on the other hand, may, in some occasions provide little information 'about a system' but could point (or bring, through learning) the observer to a set of actions that may result in higher coherence and more positive outcomes.

Complex thinking, as proposed here, builds on the tradition of systems thinking to the extent that it proposes that a set of practices of building a relation with the world, and making distinctions, may allow the enactment of particular processes leading to the creation of information with which to build a wider array of possibilities for action, leading to positive change. A variety of practices that have been developed within the systems thinking tradition are likely to make critical contributions for complex thinking (Reynolds and Holwell 2010) and that attend both to processes and contents of the thinking.

References

G. Bateson, *Mind and nature: a necessary unity* (Bantam Books, New York, 1979)

Y. Bar-Yam, *Making things work: solving complex problems in a complex world* (Knowledge Press, New England, 2004)

J. Buchler (ed.), *The philosophical writings of Peirce [originally published 1940]* (Dover Publications Inc., New York, 2014)

D. Byrne, G. Callaghan, *Complexity theory and the social sciences: the state of the art* (Routledge, London, 2014)

L. Caves, A.T. Melo, (Gardening) Gardening: a relational framework for complex thinking about complex systems, in *Narrating complexity*. ed. by R. Walsh, S. Stepney (Springer, London, 2018), pp. 149–196. https://doi.org/10.1007/978-3-319-64714-2_13

P. Checkland, J. Scholes, *Soft systems methodology in action. Includes a 30-year retrospective* (Wiley, Chichster, 1990)

R. Chia, Complex thinking: towards an oblique strategy for dealing with the complex, in *Complexity and Management*. ed. by A. McGuire, B. McKelvey (Sage, Los Angeles, 2011), pp. 182–198

P. Cilliers, *Complexity and postmodernism. Understanding complex systems* (Routledge, London, 1988)

B. Edmonds, System farming, in *Social systems engineering: the design of complexity*, ed. by C. García-Díaz, C. Olaya (Wiley, Chichester, 2018), pp. 45–64

P. Érdi, *Complexity explained* (Springer Science & Business Media, 2007)

K.T. Fann, *Peirce's theory of Abduction* (Martinus Nijhoof, The Hague, 1970)

S.O. Funtowicz, J.R. Ravetz, Uncertainty, complexity and post-normal science. Environ. Toxicol. Chem./SETAC **13**(12), 1881–1885 (1994)

R. Glanville, A (Cybernetic) musing: control, variety and addiction. Cybern. Hum. Knowing **11**(4), 85–92 (2004)

S.J.A. Kelso, *Dynamic patterns: the self-organization of brain and behavior* (MIT Press, Cambridge, MA, 1995)

J.T. Klein, Interdisciplinarity and complexity: an evolving relationship. Structure **71**, 72 (2004)

M. Lissack, A. Graber, Preface Lissack, M. & Graber, A. (Eds). *Modes of explanation. Affordances for action and prediction* (Palgrave Macmillan, New York, 2014), pp. xviii–xvi

S.M. Manson, Simplifying complexity: a review of complexity theory Geoforum. J. Phys. Hum. Reg. Geosci. **32**(3), 405–414 (2001)

H. Maturana, F. Varela, *The tree of knowledge. The biological roots of human understanding* (Shambhala, Boston, MA, 1992)

R.R. McDaniel, D. Driebe, *Uncertainty and surprise in complex systems: questions on working with the unexpected* (Springer, Berlin, 2005)

A.T. Melo, Abducting, in *Routledge handbook of interdisciplinary research methods,* ed. by C. Luria, P. Clough; M. Michael, R. Fensham, S. Lammes, A. Last, E. Uprichard (org.) (Routledge, London, 2018), pp. 90–93

A.T. Melo, L. Caves, Complex systems of knowledge integration: a pragmatic proposal for coordinating and enhancing inter/transdisciplinarity, in A. Adamsky, V. Kendon, (2020) *From astrophysics to unconventional computing: essays presented to Susan Stepney on the occasion of her 60th birthday.* Emergence, complexity, computation, vol. 35 (Springer, Cham, 2020a), pp. 337–362. https://doi.org/10.1007/978-3-030-15792-0_14

A.T. Melo, L.S.D. Caves, A. Dewitt, E. Clutton, R. Macpherson, P. Garnett, Thinking (in) Complexity: (in)definitions and (mis)conceptions. Syst. Res. Behav. Sci. **37**(1), 154–169 (2019). DOIurl: https://doi.org/doi.org/10.1002/sres.2612

E. Morin, From the concept of system to the paradigm of complexity. J. Soc. Evol. Syst. **15**(4), 371–385 (1992)

E. Morin, *Introduction à la pensée complexe* (Éditions du Seuil, Paris, 2005). [originally published in 1990]

E. Morin, Complex thinking for a complex world – about reductionism, disjunction and systemism. Systema: Connect. Matter Life Cult. Technol. **2**(1), 14–22 (2014)

J. Nubiola, Abduction or the logic of surprise. Semiotica **153**(1/4), 117–130 (2005)

M. Reynolds, S. Holwell (eds.), *Systems approaches to managing change: a practical guide* (Springer, London, 2010)

K. Richardson, The hegemony of the physical sciences: an exploration in complexity thinking. Futures **37**(7), 615–653 (2005)

K.A. Richardson, Managing complex organizations: complexity thinking and the science and art of management. Emerg.: Complex. Organ. **10**(2), 13 (2008)

K.H. Rogers, R. Luton, H. Biggs, R. Biggs, S. Blignaut, A.G. Choles, C.G. Palmer, P. Tangwe, Fostering complexity thinking in action research for change in social–ecological systems. Ecol. Soc. **18**(2), 31 (2013). https://doi.org/10.5751/ES-05330-180231

F.J. Varela, E. Thompson, E. Rosch, *The embodied mind: cognitive science and human experience* (MIT Press, Cambridge, MA, 1991)

A.N. Whitehead, *Process and reality*, Corrected edn. (The Free Press, New York, 1978)

Chapter 4
An Operational Framework for Complex Thinking

Abstract In this chapter, we present a detailed description of the framework for Complex Thinking introduced in previous chapters. We present and define a set of 9 dimensions for complex thinking and a total of 24 properties to guide practice. We discuss the current stage of development of the framework and set up a number of questions to be addressed by future investigations.

Keywords Complex thinking · Practices · Complexity · Change · Complex systems

The operationalisation of complex thinking in terms of principles and organising dimensions may be particularly useful for guiding the development of indicators and strategies aimed at describing, promoting and evaluating the complexity of the thinking, applied to the management of change in complex systems.

In this section, we introduce a preliminary framework for complex thinking, presenting a set of dimensions and properties through which it may be operationalised, with a focus on processes, rather than contents. These properties are not, in themselves, strategies or tools for intervention, albeit they advise the construction of resources to guide our thinking~acting in the world. They are necessarily embedded and part of the relational coupling practices that underlie the way we build descriptions, explanations or predictions. In our relating with the world, they may appear enacted to different degrees and with different types of expressions. Hence, the properties we present in the following sections may be used to describe the nature of our coupling practices and the extent to which they are likely to attend to and perform complexity. Their practice may support the construction of more differentiated and integrated versions~views of the world, sustaining a type of recursive, self-organised dynamics that, building coherence in that coupling, may lead to the emergence of pragmatically relevant outcomes.

Our proposal is aimed at establishing the foundations for the practice of (more) complex forms of thinking across domains. As said before, this work is open-ended and evolving. We take it as a foundation upon which to continue to build a (progressively) more robust, self-actualising (recursive) and complex (differentiated, integrated) framework for complex thinking, suited for practice in face of the world's

© The Author(s), under exclusive license to Springer Nature Switzerland AG 2020
A. Teixeira de Melo, *Performing Complexity: Building Foundations for the Practice of Complex Thinking*, SpringerBriefs in Complexity,
https://doi.org/10.1007/978-3-030-46245-1_4

challenges. This work voices a broad invitation for cross-epistemic/disciplinary interactions, explorations and applications, calling into play a variety of contexts and
practices of knowing. We hope it appeals to the critiques and contributions from others: to revise it, expand it and bring it into the realm of practice. We aim to stimulate
future dialogues about the practices supporting complex thinking, comparing differences and commonalities in different domains and promoting a cross-fertilisation
of strategies and tools. We hope to engage others in collaborative efforts aimed at
widening the horizon of possibilities for understanding, explaining and managing the
transformations in a world where we, as human observers, may act in ways leading to
positive co-evolution and cooperation, between us all and our environments. In face
of multiple challenges, at the intersection between local and global processes and at
the crossing between individuals lives and wider eco-systemic dynamics, the development of more complex modes and outcomes of thinking is an urgent necessity,
requiring collective, coordinated efforts and contributions from multiple fields.

Hence, the presentation of this preliminary framework aims at inviting others
to collaborate in its further development, engaging in comparative exercises and
exploring its applications in different domains (of living~knowing). This preliminary framework takes the shape of a list of dimensions and properties that are here
described with a focus on processes and in a content-free way. Some of the properties presented pertain more clearly to complex thinking as a process of coupling
while others can also be considered as outcomes indicators of the complexity of the
thinking. A diversity of practices can be mapped for each property, and then both
differentiated and related, in a way that is recursively congruent with the dynamics
of complex thinking.

We hope this framework guides and supports the identification of tools, resources
and practices to perform complex thinking, as a process, through the enactment of
the properties described in the next sections, and to evaluate the nature and impact
of its outcomes. It may guide systematic reviews and cross-disciplinary mappings of
existing tools and strategies, as well as the adaptation, development or evaluation of
new ones.

Some tools may enact generic processes that could be replicated for different
contents, while others could be more content and domain specific.

Notwithstanding we have been developing and exploring applications of this
framework in our own domains of action[1,2] we have deliberately avoided, in this
book, to provide examples of their practice. The reason for this choice is grounded

---

[1]Working with a focus on families, the author has been developing a Guide to Support Complex
Case Conceptualisation for the Development and Promotion of the Family's Potential for Change,
with special applications in cases of multichallenged families and child protection. She has been
developing strategies to enact the properties of Complex Thinking described in this chapter in
the process of conceptualising and dealing with family change, while attending to key "content"
dimensions for understanding Families as Complex Systems (Melo 2020).

[2]E.g. In Caves and Melo (2018) we proposed a Relational Thinking Method that has led to other
developments (Melo and Caves 2020; Melo 2020), namely its application to the facilitation of
complex thinking in Interdisciplinarity (e.g. interdisciplinary discussions or debates) and to communicating, for a general audience, the notion of relational recursive thinking (e.g. workshops at
the Festival of Ideas-York: http://yorkfestivalofideas.com/2018/community/destination-unknown/;

in the fact that providing an example would necessarily funnel the concepts and pin them to a particular reality. The future application of this framework to different domains may feedback into its development and its further differentiation and integration. At this stage, locking the framework with narrow examples could constrain the exploratory movements we wish to stimulate in different fields. By illustrating the practice of a property for complex thinking, through particular types of content, we would risk not appealing to those working with very different types of systems. On the other hand, any example would necessarily be only one of the many possible ways of enacting a particular property, even for one given domain. When a person presents or is presented with something different or new, oftentimes, they ask their interlocutors for concrete or applied examples of such a concept. But when the 'thing' at stake is relatively complex (differentiated, integrated, recursive, emergent) or fuzzy, it is most likely that one example or model will not suffice. Offering one example could deter the exploration of other facets of the phenomena, fundamental to a more holistic and thorough understanding. When, in such situations, a person does offer a particular example, others, having a first contact with the idea, may react with comments like "So, *that's* what it *is*?!". In response, and in face of the degree of complexity of the concept, the presenting person might feel compelled to respond something like: "*yes* (it is also)…but *no* (it is not only)… *and…* (there are variations, contexts, that need to be considered that might change things), *plus…* (there is also more-or less-to it, depending on the circumstances)". Because our goal with this initial presentation is to reach the widest possible audience, we will avoid pinning down the conceptual definitions of the properties of complex thinking to a particular reality. Instead, we choose to offer a more abstract, process-focused definition. We hope these properties will gain specific contours and clarity when applied to particular domains. We expect for their full apprehension and their potential for practice to be clarified in future cross-domain and cross-disciplinary dialogues and investigations. At this moment, we leave open the pragmatic exploration of the concept to a variety of domains and Systems of Interest.[3] Future work should explore a variety of possible ways of practicing (more) complex forms of thinking, identifying existing practices but also developing and evaluating new tools and strategies in a comparative way within and across domains.[4]

We invite the reader to engage in the exercise of thinking how the concepts presented may be enacted in their realities, bringing in the contents from their domains:

- What would the thinking be like (that is similar and different from how it is currently practiced)?

Sessions for schools in the context of the extension program CES Vai à Escola (https://www.ces. uc.pt/extensao/cesvaiaescola/), with the "Complex Thinking Academy workshops").

[3]For those readers who are frustrated by the lack of concrete examples we leave a note of empathy. We can understand your position. Please the contact the author if you wish to know more or get involved in ongoing work and practical applications.

[4]This study is the focus of the "Building Foundations for Complex Thinking Project"—cf. postface.

- How could different properties for complex thinking be practiced in relation to different types of problems?
- How different or similar are the current dominant modes of thinking in a particular domain and between domains?
- What difference may the nature or type of informational content of the thinking make and how can it shape or influence the process?
- How can these properties inform the way the information is produced and updated in recursive cycles of interaction with the target world?
- In what ways could the properties be (more/better) enacted? What strategies, heuristics or tools could be used?

These are some of the questions raised by the framework here presented that we hope will be addressed in future works.

In general terms, we aim to open a broad conversation based on not just one of two examples, but on a comparative exploration of a multiplicity of practices that may support complex thinking. We assume that some properties may be more salient and appear more relevant for some domains than others. Some will be equally applicable to natural, biological and physical and social systems while others will have special relevance to the latter as properties that underlie social complexity. In particular domains there may be a stronger tradition in the practice of a particular property for which a variety of existing tools and strategies may already be available, while others need to be developed.

The properties of complex thinking are thought of as constituting parts of a system of thinking, that interact non-linearly. Through their interaction, they give rise to particular types of emergent outcomes that will further constrain and shape the parts, their interaction and the unfolding dynamics of the thinking as a whole. The thinking will likely be enriched by a combination of properties. These properties might establish different types of relations with each other. These are topics to be addressed in future investigations. We postulate that a process of thinking that is organised according to such properties is more likely to generate a multiplicity of (rich) descriptions, explanations and predictions, thereby opening up more possibilities of action. Future work should map the nature of the relation between the different properties of the thinking and how different configurations could support similar or different types of emergent outcomes. For example, some properties may be important precursors of others, and some may operate as activators or hubs in the network of properties. It is necessary to further investigate how they organise themselves in relation to each other, how they interact and under which conditions they support particular types of outcomes. Hence, it is necessary to map the relational structure and dynamics of complex thinking by mapping the structure and dynamics of the relations between the properties that enact it. Different configurations of properties, organised both synchronically and diachronically, may also be more or less critical to support positive outcomes for particular types of Systems of Interest.

Different properties of the thinking will generate different types of information and perspectives/lenses that will lead to the differentiation of the thinking in different dimensions. The thinking may evolve through processes of iteration and recursion of thinking trajectories where different variations and ways of integration are explored, generating new perspectives (Caves and Melo 2018). The notion of a *thinking trajectory* refers to a process that experiments with a particular configuration and sequence of properties, and that unfolds in a particular way. Some properties may be practiced sequentially, while some may be embedded in others, while some may create a context or stage for others to flourish; some tools to support/promote complex thinking may target a specific (and say smaller) set of properties, while other tools may need to be combined, in order to target a wider set. As experience of promoting complex thinking grows, new tools are likely to emerge that employ a variety of strategies for orchestrating a network of interrelated properties in planning for and managing the thinking process. It is necessary to identify properties for which there are insufficient tools or resources and to stimulate the development of new ones.

In certain circumstances, the rehearsal of the interaction between the lenses generated by the practice of different properties, may lead to the emergence of a higher degree of novelty in the form of hypotheses that can inform new actions and produce additional information that further contributes to the complexity of the thinking. The intentional management of this process may imply the skilful use of a multiplicity of strategies and tools and, above all, a way of coordination and integrating them.

Our proposal may guide future research to build the necessary knowledge that leads to further development of an operational framework for complex thinking into: (i) a set of micro-frameworks (applied to particular domains); (ii) a meta-framework of principles that guides the performance of complex thinking through the intentional orchestration and coordination between different properties, in building trajectories of thinking that increase its (relative) complexity, as it unfolds; (iii) a meta-methodological approach that guides the choice of tools and strategies to support the practice of complex thinking. The development of such meta-methodological approach may be realised as an abducting approach (Melo 2018) by: (i) establishing a set of practices and strategies for the development of complex thinking as a form of abductive reasoning; (ii) nurturing a basic inquisitive stance and a strong, attuned and sensitive contributions for the coupling with the target system-of-interest; (iii) proposing a set of strategies for the practice of complex thinking generating rich information and a multiplicity of perspectives about the target system of interest as well as the coordination and creative exploration of how these perspectives can be combined to lead to novel, emergent insights. When these processes occur in the context of interdisciplinary interactions, they become strategies for interdisciplinary abducting (Melo 2018) for the recursive development of complex thinking, through interdisciplinarity and the development of complex thinking for (richer or novel) interdisciplinarity.

In the following sections, we offer a tentative definition of each dimension and associated properties of complex thinking that integrate our proposal into an operational framework for complex thinking. We will attempt to provide a process-focused, content-free definition of each property. There are many properties identified or associated with the operations of biological and natural systems as well as in many social and human systems. This list results from our exploration and integration of the literature in different domains in relation to complexity, but it is not an exercise of listing properties of complex systems. We do not expect these properties, when applied to the thinking, to necessarily behave or be expressed in a direct correspondence with what may be seen in other complex systems. They should be seen more as heuristics for discovery. The challenge for future work will be to identify particular ways in which these properties may be enacted at the level of our thinking. In the future, it will be relevant to explore the literature around these concepts, mapping existing contributions and how they relate to each other and to practice. For that reason, and for the sake of simplicity of presentation, we do not provide specific literature references for each property. For the reader unfamiliar with some of these concepts we recommend an exploration of the literature for examples of the corresponding properties as expressions of the complexity of different types of system.

We believe that the enactment of these principles, corresponding to particular processes, may be especially relevant to the understanding and/or management of change in (relatively) complex systems, through first and second-order complex thinking. On the one hand, this list represents an attempt ata full characterisation of the thinking processes or their complexity; on the other, it should remain open to transformations (as a result of being subject to complex thinking).

4.1   Structural Complexity

Structural complexity relates to the nature of the elements that organise the thinking and the nature of their relation to each other. It also pertains, to some extent, to the contents of the thinking and how much they attend to known complex properties of the world (e.g. it may include contents of 'complexity thinking'[5]). Nevertheless, it is mostly defined by the nature or kind of these contents and their relation to each other. It also pertains to the nature of the basic movements performed in the thinking in terms of how the information is explored, manipulated and created.

4.1.1   Structural Variety and Dimensionality

This property relates to the extent to which the thinking enacts and results in a variety of acts of distinctions and indications and ways of constructing multiple perspectives on the target systems of interest. It relates to the extent that the thinking includes a variety of elements and information of different kinds that is produced in the context of a strong coupling that allows for the information to be expanded, enhanced and enriched. Additionally, it pertains to the extent to which thinking includes/creates

---

[5]See Sect. 3.2.

information corresponding to multiple dimensions and subdimensions of the relational world of the target system.[6] This property pertains to the differentiation of the thinking.

4.1.2 Relationality

Relationality pertains to the extent to which the thinking shows an organisation that allows for the exploration of the information in relational terms, where each bit of information or dimension is explored in the context of their relation to each other, considering multiple types of relations or relational properties. It also encompasses: (i) the extent to which the thinking explores a variety of kinds of relations and properties between the different elements of the thinking and the information pertaining to the target system of interest; and (ii) the extent to which diverse elements or dimensions that are part of the thinking are integrated.

4.1.3 Recursiveness

The property of recursiveness refers to the extent to which the thinking (i) creates and explores the available information with a multiplicity of movements that are non-linear (different movements relate different elements or dimensions of the thinking creating and exploring different configurations of relations) and (ii) includes recursive turns, hence allowing for the same, or previously explored dimensions to be revisited to allow for the exploration of new relations, thereby transporting and integrating information from subsequent points in the thinking trajectory to (re)visit previously explored dimensions and relations.

This property also pertains to the extent to which the thinking trajectory builds circular itineraries in the exploration of the different dimensions of the target system and its relational world. Higher order patterns or organising concepts can be identified through the exploration of circular trajectories, supported by core configurations of relations that, through recursive loops, stabilise a certain number of dimensions in a network of relations.

4.2 Dynamic/Process Complexity

Dynamic/process complexity refers to the organisation of the thinking in relation to time and the way the thinking manages its own dynamics and processes while attending to those of the world.

---

[6]Cf. Caves and Melo (2018).

4.2.1   Timescales

The thinking might be characterised by the way it unfolds in multiple time-scales and temporal modes that are both differentiated and integrated and interact in ways that generate multiple differences resulting in rich information and multiple perspectives. The extent to which the thinking, through its unfolding recursiveness, informs its own transformation and avoids being locked in one timescale and temporal mode, keeping open the possibility of exploring multiple times and time-scales. The extent to which the thinking is capable of attending and attuning to multiple timescales and time modes of the systems of interest.

4.2.2   Dynamic Processes

This property pertains to the extent to which the thinking has a process orientation and is organised dynamically, generating a succession of differences (e.g. between dimensions, moments, contexts) and variations of information, in multiple dimensions, from which multiple perspectives emerge. The extent to which the thinking is focused on processes (more than contents or static states) that connect different elements and events of a relational world as well as their (coordinated) transformations.

4.2.3   Relativity, Ambiguity and Uncertainty

The thinking may be characterised by the extent to which it is embedded in a relative stance that results from an explicit or implicit consideration of a relational world. This property relates to the extent that the thinking frames and contextualises its descriptions, explanations and previsions acknowledging a particular relative position. It also relates to how the thinking enacts tolerance and acceptance for a multiplicity of (relative) perspectives and how it embraces and integrates the ambiguity or uncertainty that result from the complexity of the world, namely its relationality and relativity, remaining open to a variety of relations with other different positions.

4.3   Causal and Explanatory Complexity

Causal and explanatory complexity refers to how the thinking addresses causality and explanation and, through them, opens possibilities for action.

4.3.1 Complementary Modes of Explanation and Finalities

This property pertains to the extent that the thinking is driven by particular finalities and selects different modes of organisation (e.g. of a descriptive, explanatory, predictive or speculative nature) that are congruent with them. The extent to which the thinking acknowledges the potentialities and limitations of its organising modes, alternating or integrating distinct and complementary modes.

4.3.2 Historicity

This property pertains to the extent that the thinking attends to its own path-dependency, preserving the memory of past trajectories which includes the history of its structural arrangements, movements and transformations, points of choice, bifurcations and changes of states along with the contextual conditions that constrained and the memory of its outcomes.

4.3.3 Complex Circularity: Part-Whole Relations

This property pertains to the extent to which the thinking is organised in circular terms, performing distinctions that identify multiple levels and considering at the level of any target emergent-whole/phenomena in relation to the level of its constituent parts/elements and to the level of its boundary and contextual conditions, exploring them, circurlalry, in terms of their mutual differences and interdependencies. The extent to which the thinking identifies/constructs and moves across new and/or different levels attending as much to the characteristics of each level as to the inter-level processes that connect those levels and mutually shape them. The extent to which the thinking expands or collapses levels and explores different properties of their inter-relations according to the extent that they provide access to information that either expands and enriches or diminishes and impoverishes the possibilities for action. The extent to which the thinking manages its own constraints (e.g. experimenting with different trajectories and boundary conditions) and explores the properties of the relations and different types of relations between its own parts (initial structural elements) and its emergent outcomes to create or facilitate the emergence of new information or generate new perspectives.

4.3.4 Emergence

The extent to which the thinking is organised in a way that leads to emergent outcomes in terms of new elements or dimensions that were not previously available to the thinking, and that cannot be directly reduced to those previously known elements (parts). The extent to which the thinking leads to novel descriptions, explanations, predictions, in the form of hypotheses, that inform new courses of action in the context of which they can be tested and/or lead to the emergence of new information. The extent to which the emergent outcomes of the thinking are used as constraints under which the previously available information can be re-shaped, re-interpreted, explored and related in new ways further contributing to new insights as well as clarifying further informational needs and demands.

4.4 Dialogic Complexity

Dialogic complexity refers to the way the thinking integrates and deals with the (seemingly) contradictory, opposing or complementary aspects of the world and with the processes that support them.

4.4.1 Dualities and Complementarities

The extent to which the thinking organises dualities and contradictions in terms of dynamic complementarities, recognising the mutually defining nature of opposites and dualities in terms of complementary pairs, integrating or bridging them. The extent to which the thinking considers its own structural components in terms of flexible complementary pairs, that it defines in relation to its limits, potentialities, conditions of applicability and respective consequences. The extent to which the thinking avoids rigid dualistic or binary positions and establishes frames of meaning that allow for a flexible relation between complementary pairs and for changes in the nature of their relation as well as for positive courses of action in relation to the target system-of-interest.

4.4.2 Trinities and Levels

The extent to which the thinking considers the processes that generate its own dualities and complementarities and those of the targeted world and attends to underlying processes that: (i) at lower level(s) explain the emergence of the duality/complementarity and (ii) at higher level(s), integrates it, eventually giving rise to a novel complementarity. The extent to which the thinking alternates and relates different levels in order to avoid or integrate paradoxes or transform rigid dualities into (a web of) eventually new, flexible complementary pairs. The extent to which the thinking identifies the types of actions that are more congruent with the nature of each level and with each choice of framing in terms of the trinities (the dualities and the processes that explain them) and their organising levels.

4.5 Observer's Complexity

The observer's complexity relates to the extent that the observer makes specific active contributions towards the complexity of the thinking through the management of their own positioning in the coupling relation.

4.5.1 Multipositioning

This property relates to the extent that the observer experiments with their positions in relation to the target system-of-interest and their environment, and manages their particular contribution to the coupling relation in ways that generate a variety of information upon which different perspectives or views on the target system can be created, and multiple acts of distinctions and indications can be performed, towards a complex understanding about the system or situation of interest.

4.5.2 Reflexivity

Reflexivity pertains to the extent that the observer turns the thinking process towards itself and their own role in particular configurations of relations related to the process of thinking itself, as well as the observer's internal factors (e.g. emotional, physical and cognitive factors; values, habits, intentions, preferences) and external contextual constraints (e.g. social context, physical environmental conditions) that shape the course, nature and outcome of the thinking. It also refers to the extent that the observer has a rich view about the potentialities and limitations of the thinking.

4.5.3 Intentionalities

Intentionalities pertain to the way that the observer's intentions, preferred visions, aspirations, values and goals drive the thinking process and to the extent that the observer identifies the constraints posed by their intentions. It also pertains to the extent that the thinking process is managed purposefully and that there is coherence between the observer's intentionalities and the thinking process and outcome.

4.6   Developmental and Adaptive Complexity

Developmental and complexity refers to the extent that the thinking develops towards increasing differentiation and integration and capacity for emergence, being capable of adapting and co-evolving coherently with its own transformations and those in the target system of interest, their environments and the configurations of relations which embed them.

4.6.1   Developmental Adaptive Value

This property relates to the extent that the thinking results in courses of action that are eco-systemically fit and sustainable for a given outcome or purpose, and the extent to which they support the positive co-evolution of the target system-of-interest and their critical observers, as well as their embedding environments, without significant harm or negative unexpected consequences.

4.6.2   Developmental Evolvability

This property pertains to the extent to each the thinking develops towards increasing complexity by creating and integrating means that support its own transformation in relation to transformations in its environment, in the target system-of-interest and in the configurations of relations that sustain them. It relates to the extent that the thinking is capable of identifying positive and negative constraints to its own evolution and implement strategies that facilitate a positive development towards increased complexity.

4.7   Pragmatic Complexity

Pragmatic complexity refers to the extent that the thinking leads to a variety of possibilities for action and informs the choice of pragmatically relevant courses of action that result in positive and sustainable or adaptive outcomes, as recognised by a multiplicity of critical observers at different moments in time.

4.7.1 Pragmatic Value

This property relates to the extent to which the thinking results in a type of novelty that appears revealing/insightful and opens pragmatically relevant new courses of action. It pertains to the extent that the thinking results in a sense of augmented understanding and in an increased number of pragmatically viable possibilities for action, and the extent to which it leads to choices likely to support positive outcomes, for given purposes, as recognised by a multiplicity of critical observers.

4.7.2 Pragmatic Sustainability

The extent to which the pragmatic value of the thinking and the courses of action to which it leads are sustainable in terms of their positive outcomes that are capable of being preserved or transformed in positive ways, while avoiding unwanted consequences or harm, perceived to a multiplicity of critical observers, within given timescales.

4.8 Ethical and Aesthetical Complexity

Ethical and aesthetical complexity refer to the extent that the process and the outcome are congruent with the ethical and aesthetical stance that motivates them and results in both good and beauty that are recognised by a multiplicity of critical observers.

4.8.1 Ethical Value

The extent to which the values that guide the process of the thinking, and that underlie its outcomes, are made explicit and congruent with the ethical frame of reference of the observer, and that are recognised as valuable and considered as Good by a multiplicity of critical observers.

4.8.2 Aesthetical Value

The extent to which the process and outcome of the thinking result generate a sense of coherence, clarity, emergent and integrative harmony and Beauty as recognised by a multiplicity of critical observers.

4.9   Narrative Complexity

The extent to which the thinking is organised, or results in, a narrative that sustains a multifaceted reality which, at some level, multiple critical observers can acknowledge or recognise, and that is flexible, differentiated and integrated, and capable of evolving to accommodate emerging novelty.

4.9.1   Differentiation and Integration

The extent to which the process and outcome narrative that supports and/or expresses the thinking is differentiated and integrated and capable of holding a variety of dimensions and sub-dimensions while sustaining a higher-level integration that generates a sense of coherence, amidst diversity and multidimensionality that gives it meaning to all its elements. The extent to which the narrative is capable of expressing the complex properties of the thinking and of synthesising, in a highly differentiated, integrated and recursive way, a variety of perspectives, timelines and timescales, contexts, recursive loops and mini-narratives, into a coherent whole.

4.9.2   Identities

The extent to which the narrative that supports and/or expresses the thinking is organised in a way that, independently of its multidimensionality and the variety of perspectives it integrates, allows for a multiplicity of critical observers to recognise their preferred identity(ies). The extent to which the narrative offers, to multiple critical observers, a variety of possibilities for action that may correspond to more positive or preferred identities, in relation to themselves and each other.

4.9.3   Flexibility/Openness

The extent to which the narrative contains elements of novelty, flexibility and openness that allow it to be transformed and evolve, avoiding rigid states or final closures.

References

L. Caves, A.T. Melo, (Gardening) Gardening: a relational framework for complex thinking about complex systems, in *Narrating Complexity*, ed. by R. Walsh, S. Stepney (Springer, London, 2018), pp. 149–196. https://doi.org/10.1007/978-3-319-64714-2_13

A.T. Melo, in *Routledge Handbook of Interdisciplinary Research* Methods, ed. by C. Luria, P. Clough, M. Michael, R. Fensham, S. Lammes, A. Last, E. Uprichard (org.) (Routledge, London, 2018), pp. 90–93

A.T. Melo, Método de pensamento relacional complexo para facilitação da emergência e integração de ideias em debates, tertúlias e outros encontros dialógicos. v2.pt.2020 (2020). https://doi.org/10.13140/rg.2.2.21504.38409

A.T. Melo, L.S. Caves, Relational thinking for emergence: a methodology for guided discussions. v1.2019 (2020). https://doi.org/10.13140/rg.2.2.30469.70881

Chapter 5
Discussion: From Thinking to Performing Complexity

Abstract The challenges of thinking complexity have led to the search for progressively more sophisticated models, simulations and other techniques that aim both at fundamental understanding and also at increasing our capability for better management of the world and its change processes. However, both the natural and the human social worlds reveal a pathway to complexity that has been underexplored: one where we cannot just learn about complexity, but need to learn to *perform* complexity. In this chapter, we review the grounds of our proposal of a new pragmatically-oriented theoretical framework for the practice of complex thinking. We reflect on how our proposal may address Morin's call for a more "Generalised" approach to complexity, in contrast to modes of "Restricted Complexity". We discuss the potential of our proposed definition and framework for Complex Thinking and its implications for future research and practice.

Keywords Complex thinking · Performing complexity · Generalised complexity · Practice

5.1 From Thinking to Performing Complexity

The challenges of thinking complexity have led to the search for progressively more sophisticated models, simulations and other techniques that aim both at fundamental understanding and also at increasing our capability for better management of the world and its change processes. But both the natural and the human social worlds reveal a pathway to complexity that has been underexplored: one where we can not just learn about complexity, but need to learn to *perform* complexity. The complex world, including the one of our own minds, reveals a set of properties that supports amazing outcomes. They allow for continual differentiation, integration and recursive development, leading to and resulting from the emergence of new configurations of relations that open new dimensions (and thus possibilities) in a changing world. By learning how to enact (at least some of) the same properties that we have identified in natural worlds as well as in the human social human realms, we may find ways to establish a different relation with the world: one which opens new possibilities for actions and guides us into potentially more positive and pragmatically relevant choices, even in possession of only partial or limited information. By learning to perform complexity, through the enactment of the properties of complex thinking, we may find ways to move and act in the world that are capable of creating positive, meaningful and sustainable changes.

Morin's call for complex thinking has yet to be fully addressed. The Sciences of Complexity are still, to a large extent, operating within the scope of a "restricted" approach to complexity (Morin 2007; Melo et al. 2019). Hence, it is necessary to create new spaces for the development of a more "generalised" approach (Ibid) and to develop a rich repertoire of practices to perform complexity along with a choreographic framework to coordinate them. We have presented a proposal for an operational framework for complex thinking, which may propel the development of such a repertoire, targeted at different properties of complex thinking. The future development of the framework should explore key coordinated movements that build synergies between them. This proposal is grounded in a relational and pragmatic worldview from which we derived basic ontological and epistemological foundations and implications for addressing complex thinking as a relational property. We present complex thinking both as a mode or process of coupling and as an outcome, to which an observer may make particular contributions. We also present complex thinking as a concept to be understood in relative terms and as a dynamic concept, likely to evolve. We have compiled a list of dimensions and properties that we currently believe to be relevant for the practice of complex thinking, as a mode of coupling congruent with the complexity of the world that has the potential for leading to more positive outcomes. The proposed framework of dimensions and properties of complex thinking supports the systematic identification, development and evaluation of tools, strategies and heuristics to support their enactment, and to new strategies for their integration. This operational definition of complex thinking is aimed at bringing complex thinking from the theoretical realm into practice. Traditions such as systems thinking, second-order cybernetics and others, have long made significant contributions to increase the complexity of our thinking. Nevertheless, we believe that some important properties of the thinking have been either neglected, insufficiently developed in practice, or have a minimal expression in some domains. On the other hand, (more) complex thinking and emergent complex thinking requires a diverse set of integrated practices that, due to enacting different properties, need to be purposefully coordinated. The proposed framework sets the foundations for future work, exploring the nature of the relation between different properties and the consideration and how they can be recursively applied to the process of coupling itself, to increase the complexity of the thinking. More complex approaches need to be developed, aimed not just at understanding and managing the complexity of the world, but also, in a close recursive relationship, the understanding and management of the complexity of the thinking itself.

This framework invites the future development of a meta-methodological approach for the intentional management of the complexity of the thinking, through exploring the relation between different properties and the practices that enact them; it is time for research and practice to strengthen their joint efforts in exploring the nature of the relations between different properties of complex thinking and how their network of relations could be intentionally managed to lead to more complex emergent outcomes.

This is also a call for an interdisciplinary approach to complex thinking, since the tools and strategies that are currently used in one particular domain could, eventually, be adapted or stimulate the development of new approaches in other domains. In a complementary manner, complex thinking can guide the coordination of the processes of interdisciplinary interactions, particularly when applied to real world complex challenges.

The framework here presented needs to be further developed into practice. It is necessary to identify existing tools and strategies in different domains that appear to address different properties of complex thinking. A mapping of existing tools and resources will shape the development of new tools and evaluation strategies that target specific properties, or their interactions with others. Additionally it is necessary to define, for specific domains, indicators of the complexity of the thinking, for which particular assessment strategies may be devised.

We hope this proposal invites new dialogues, within and between disciplines, capable of strengthening both the theoretical foundations and the structure of a pragmatic framework to guide complex thinking in application to the understanding and management of change in complex systems. We believe that the concept of complex thinking will continue to be developed both at a theoretical and operational level. We anticipate that other properties may be added, and some may be combined, and we foresee the development of a range of practices and tools.

The biggest challenges rests in the pragmatic test of the concept, particularly of second order complex thinking, in specific domains. Therefore, we invite academics, researchers and practitioners to join efforts in constructing conditions for the development and evaluation of our practices of coupling with the world, towards increased complexity. It will be the careful investigation and supported practice that will dictate to what extent, and under which conditions, new integrated practices of complex thinking may result in differences that make a difference (Bateson 1979).

Finally, we believe that a relational worldview, such as the one espoused, is capable of bridging both realist and constructivist arguments and creating a space where they can be conciliated in pragmatic terms. This would free the concept of complex thinking from debates where its operational development would be blocked by different ontologies and epistemologies. The framework presented here is also a bridging tool between Science 1 and Science 2 (Lissack, and Graber 2014) and between 1st and 2nd second order science and cybernetics (Riegler et al. 2018), opening new possibilities for more complex constructions of the world and a more complex knowledge construction and practice.

References

Bateson, G, in *Mind and Nature: A necessary unity*. (New York, Bantam Books, 1979)

M. Lissack, A. Graber, Preface, in *Modes of Explanation. Affordances for Action and Prediction*, eds. by M. Lissack, and A. Graber, A, (New York, Palgrave Macmillan, 2014) pp. xviii–xvi

E. Morin, Restricted complexity, general complexity, in *Worldviews, Science and Us. Philosophy and Complexity*, eds. by E. Gersherson, D. Aerts, B. Edmonds, (London, World Scientific) pp. 5–29

A.T. Melo, L.S.D. Caves, A. Dewitt, E, Clutton, R. Macpherson, P. Garnett, Thinking (in) Complexity: (In)definitions and (mis)conceptions Syst. Res. Behav. Sci. **37**(1), 154–169 (2019). https://doi.org/10.1002/sres.2612

A. Riegler, K.H. Muller, S.A. Umpleby (eds.), *New Horizons For Second-order Cybernetics* (World Scientific, Singapore, 2018)

Postface

This book introduces the foundations for a framework for the practice of complex thinking applied to the management of change in complex systems. This framework can provide guidance for a cross-disciplinary investigation of the practice of complex thinking in different domains and an interdisciplinary approach to the development, selection, and evaluation of strategies and tools for complex thinking applied to the management of change in complex systems. This study is the focus of the "Building Foundations for Complex Thinking Project", led by the author, Ana Teixeira de Melo (at the Centre for Social Studies of the University of Coimbra, Portugal) and co-led by Leo Caves (independent researcher, Associate of the York Cross-Disciplinary Centre for Systems Analysis of the University of York, United Kingdom and collaborator of the Centre for the Philosophy of Sciences of the University of Lisbon, Portugal) and Philip Garnett (York Management School, University of York). This project will also attempt to develop content-free templates of tools to support the practice of specific properties of complex thinking as well as to guide the choice of tools; these tools are also developed for specific domains. This project will further develop the framework to explore the structure and dynamics of the relations between properties of complex thinking and how they can be intentionally managed, during the planning, implementation and evaluation of an intervention—in this complex world—in order to support positive outcomes.

Please contact the author if you are interested in getting involved.

69

A. Teixeira de Melo, *Performing Complexity: Building Foundations for the Practice of Complex Thinking*, SpringerBriefs in Complexity,
https://doi.org/10.1007/978-3-030-46245-1

Printed in the United States
By Bookmasters